Applying Predictive Analytics

Richard V. McCarthy • Mary M. McCarthy
Wendy Ceccucci

Applying Predictive Analytics

Finding Value in Data

Second Edition

 Springer

Richard V. McCarthy
Quinnipiac University
Hamden, CT, USA

Mary M. McCarthy
Central Connecticut State University
New Britain, CT, USA

Wendy Ceccucci
Quinnipiac University
Hamden, CT, USA

The online files of this book are available in the https://link.springer.com/book/10.1007/978-3-030-83069-4

ISBN 978-3-030-83072-4   ISBN 978-3-030-83070-0   (eBook)
https://doi.org/10.1007/978-3-030-83070-0

This Springer imprint is published by the registered company Springer Nature Switzerland AG
The registered company address is: Gewerbestrasse 11, 6330 Cham, Switzerland

This book is dedicated to:
Camryn Fraello, Kylie Fraello, Jack Cantor,
Owen McCarthy, Cora Cantor, and Shane
McCarthy
To my amazing family, with all my love and
gratitude.

Preface

This book is intended to be used by students who are taking a course in predictive analytics within an undergraduate or graduate business analytics program. The book uses extensive examples and describes how to use the SAS Enterprise Miner™ software. SAS Enterprise Miner™ is one of the world's leading analytics software. For instructors, SAS offers a university academic alliance program for educational programs as well as a joint certification program for your university.

This book begins by introducing analytics along with a model for the use of predictive analytics to analyze business problems and opportunities. In the information age we now live in, we have the capability to analyze vast amounts of data. Chapter 2 focuses on how to use SAS Enterpise Miner™ to evaluate and prepare the data for analysis and create a predictive model.

Although it is assumed that the students will have a basic understanding of business statistics prior to taking a course in predictive analytics, Chap. 3 presents a comprehensive review of business statistics that are necessary for analyzing the results of the predictive analytic models that are presented in the remaining chapters. Chapters 4, 5, and 6 present what are considered the big three in predictive analytics: regression, decision trees, and Neural Networks. Chapter 7 presents several predictive analytic methods that are relatively new such as random forests, boosting, and ensemble models. We conclude with an explanation of how to use predictive analytics on current data to take advantage of new business opportunities.

The focus of the book includes not only explaining the predictive analytics techniques but demonstrating how to use them utilizing the SAS Enterprise Miner™ software, and how to interpret the output to make business decisions. This is beneficial to data scientists, business analysts, financial analysts, systems analysts, and others who are involved in analyzing large volumes of data.

New to the second edition, we have expanded the coverage on regression to include other forms of regression, not supported directly by SAS Enterprise Miner. However, these may get support through the use of statistical software such as Python, R, or SAS. Two new chapters were added. Chapter 8 provides a detailed description on the use of cluster analysis. This is particularly useful in data exploration for predictive analytics, when the analyst is uncertain which variables to utilize. Chapter 9 provides a detailed description of the use of text mining as a means to augment a predictive analytics data model by incorporating qualitative data.

For instructors who have adopted this textbook, you will note that throughout the book we utilize a dataset dealing with automobile insurance claim fraud as an example to illustrate each of the predictive analytics techniques. There are two data sets. The first is the data set that includes the historical data for analysis and the second is the scoring data set. The data sets are available for instructors only, in either a SAS or CSV format, along with an XML diagram of the complete example. In addition, new to the second edition, we have included a case in Appendix B&C dealing with analysis of banking transactions that potentially involve money laundering. This could serve as an additional example within a course or as a semester project. There are two data sets that accompany this case. The first is the historical data for analysis and the second is the scoring data set. The data sets are available in CSV format and instructors' notes are available upon request.

Hamden, CT, USA   Richard V. McCarthy

Contents

About the Authors

Richard V. McCarthy (DBA, Nova Southeastern University, MBA, Western New England College) is a professor of computer information systems in the School of Business, Quinnipiac University. Prior to this, Dr. McCarthy was an associate professor of management information systems at Central Connecticut State University. He has 20 years of experience within the insurance industry and has held a charter property casualty underwriter (CPCU) designation since 1991. He has authored numerous research articles and contributed to several textbooks. He has served as the associate dean of the School of Business, the MBA director, and the director of the Master of Science in Business Analytics program. In 2019, he was awarded the Computer Educator of the Year from the International Association for Computer Information Systems.

Mary M. McCarthy (DBA, Nova Southeastern University, MBA, University of Connecticut) is a professor and chair of accounting, Central Connecticut State University. She has 20 years of financial reporting experience and has served as the controller for a Fortune 50 industry organization. She holds a CPA and CFA designation. She has authored numerous research articles.

Wendy Ceccucci (PhD and MBA, Virginia Polytechnic University) is a professor and chair of computer information systems at Quinnipiac University. Her teaching areas include business analytics and programming. She is the past president of the Education Special Interest Group (EDSIG) of the Association for Information Technology Professionals (AITP) and past associate editor of the *Information Systems Education Journal* (ISEDJ). Her research interests include information systems pedagogy.

Chapter 1
Introduction to Predictive Analytics

"It is a capital mistake to theorize before one has data. Insensibly one begins to twist facts
to suit theories, instead of theories to suit facts."
　Sir Arthur Conan Doyle

"In God we trust, all others must bring data."
　W. Edwards Deming

"Know where to find information and how to use it; that is the secret to success."
　Albert Einstein

Chapter Learning Objectives
1. Define business analytics, big data, and predictive analytics.
2. Differentiate descriptive analytics from predictive analytics.
3. Describe the nine-step predictive analytics model.
4. Explain why a critical thinking mindset is necessary when solving business
   problems.

There are few technologies that have the ability to revolutionize how business
operates. Predictive analytics is one of those technologies. Predictive analytics consists primarily of the "Big 3" techniques: regression analysis, decision trees, and
neural networks. Several other techniques, such as random forests and ensemble
models, have become increasingly popular in their use, predictive analytics focuses
on building and evaluating predictive models resulting in key output fit statistics that
are used to solve business problems. This chapter defines essential analytics termi-
nology, walks through the nine-step process for building predictive analytics models, introduces the "Big 3" techniques, and discusses careers within business
analytics.

An example using automobile claim fraud identification in the property-casualty
insurance industry is used throughout this book to illustrate each concept and
technique. Appendix A contains a detailed description of the metadata for this
example.

R. V. McCarthy et al., *Applying Predictive Analytics*,
https://doi.org/10.1007/978-3-030-83070-0_1

1.1   Predictive Analytics in Action

In an increasingly fierce competitive market, companies need to identify techniques that garner competitive advantages to thrive. Imagine a company with the ability to predict consumer behaviors, prevent fraud, mitigate risk, identify new customers, and improve operations. What if in real time an organization can:

- Identify customer's spending behaviors and cross-sell efficiently or sell additional products to their customers.
- Enhance customer satisfaction and customer loyalty.
- Identify the most effective marketing campaign and communication channels.
- Identify fraudulent payment transactions.
- Flag potential fraudulent claims and pay legitimate insurance claims immediately.
- Predict when machinery will fail.

With the advancement in computer hardware (faster processing speeds, in-line memory, cheaper storage, and massively parallel processing (MPP) architectures) coupled with new technologies such as Hadoop, MapReduce, Hive, Pig, Spark, AWS, Azure, and MongoDB and analytics for processing data, companies are now positioned to collect and analyze an enormous amount of structured and unstructured data gaining valuable insights from the data in real time (run predictive algorithms on streaming data) or near real time.

Technology has enabled analytics. The ability to manage and use a massive amount of data and advances in speed and memory capacities have now made it possible to easily implement processes such as neural networks, which have existed primarily in theory for decades.

Amazon, the number 1 online retailer, uses predictive analytics for targeted marketing. Their algorithms analyze seemingly endless customer transactions searching for hidden purchasing patterns and relationships among products, customers, and purchases. Their collected data includes purchase transactions, information that is contained in their customers' wish and shopping lists, and products the customers reviewed and searched for the most. Their personalized recommendation systems, "Customers who bought this item, also bought this one…" and "Frequently bought together," are based on effective predictive analytics algorithms. Further, Amazon patented a predictive algorithm that anticipates what customers are going to buy and then ship goods to an Amazon warehouse nearby prior to the customer buying the goods. Then when the customer orders the product, having anticipated the order, Amazon can ship very quickly (Kopalle 2014).

Netflix also has a recommendation system ("You might want to watch this…") that uses member's past viewing behavior to recommend new movies or shows. Netflix was able to predict that "House of Cards" would be a hit before the show went into production. Using predictive analytics, Netflix determined that a series directed by David Fincher, starring a certain male actor, and based on a popular British series was worth a $100 million investment (Carr 2013). Netflix gathers a huge amount of data from their 130 million memberships. Netflix collects what the members are streaming, the ratings given by their members, their searches, what day of the week, how many hours per month they watch shows, as well as what members do when the credits roll. With all of this data, Netflix can see trends and membership behavior and make predictions about what new streaming content to offer, personalized recommendations to make with the intent of increasing and retaining memberships. They identified that customers "binge-watch" and then used that to differentiate their streaming platform.

Although this is now a maturing technology, it has helped shape expectations among non-technical business leaders. Gartner reports that because of advances in analytics and business intelligence platforms there are often expectations of artificial intelligence-driven data stories created by insights generated from ongoing monitoring of relevant data. This may include user feedback, ratings, and usage behavior (Richardson et al. 2021).

Walmart, the number 1 retail giant, uses predictive analytics to determine which products to offer in stores that are in a hurricane's path. Based on historical data purchasing behavior from Hurricane Charley, they were able to predict that stores would need not only flashlights but also strawberry Pop-Tarts. They noticed that strawberry Pop-Tarts sales increased seven times their typical rate. Based on this knowledge, Walmart was able to stock the stores in the path of Hurricane Frances with the desired products. The products sold, and Walmart was able to profit from their predictive analytics (Hays 2004). Predictive analytics is a vital part of Walmart's strategy. Walmart uses analytics to predict stores' checkout demand at certain hours to determine how many associates are needed at the checkout counters. Through their predictive analytics, Walmart is able to improve the customer experience and establish the best forms of checkout, i.e., self-checkout or facilitated checkout by store. Similarly, Walmart uses predictive analytics in their pharmacies by analyzing how many prescriptions are filled in a day and the busiest times during a day and month and then optimizes their staffing and scheduling to ultimately reduce the amount of time it takes to get a prescription filled. Additionally, Walmart uses predictive analytics to fast-track decisions on how to stock store shelves, display merchandise and new products, and discontinue products and which brands to stock (Walmart Staff 2017). Walmart introduced a social shopping app named Shopycat during the holidays. The Shopycat app analyzes data on Facebook and provides Facebook users the ability to find gifts for friends and family based on their tastes and interests. Recently, Walmart hired Ryan, a 6-year-old YouTube sensation known for Ryan ToysReview channel. His six YouTube channels draw more than a billion views each month (Reuters 2018). Walmart likely used predictive analytics to create his toy and apparel offerings, and the launch of those products would be a success. Like Target, Walmart collects an enormous amount of data on consumer behavior and demographics. Predictive analytics is an essential part of their strategy in understanding consumer behavioral patterns and trends.

A Harley-Davidson dealership in the New York City area used predictive analytics to optimize their marketing campaign resulting in increased leads and sales as well as defining a larger customer base. The predictive analytics used headlines, visuals, key performance targets, as well as defining characteristics and behaviors of high-value past customers. These characteristics and behaviors included customers who had completed a purchase, added a product to the dealership's online shopping cart, viewed the dealership's websites, or were among the top 25% in most time spent on the website. Using predictive analytics, the dealership was able to identify potential customers ("lookalikes") who resembled past customers. The analysis predicted which possible headlines and visual combinations were most effective. For example, the analysis revealed that ads with the word "call" performed 447% better than ads with the word "buy" (Power 2017).

The retail industry is certainly well-entrenched in the use of predictive analytics. Other organizations also use predictive analytics. UPS, a global leader in logistics, uses the data they collect to optimize their logistics network. Launched in 2013 and upgraded in 2017, their On-Road Integrated Optimization and Navigation (ORION) system, their fleet management system, uses predictive analytics techniques to optimize their vehicle delivery routes dynamically. UPS' chief information and engineering officer, Juan Perez stated "… it will lead to a reduction of about 100 million delivery miles and 100,000 metric tons of carbon emissions." Their use of advanced analytics has led to a greater efficiency in operations, fuel and vehicle savings, and a reduction in greenhouse gasses (Samuels 2017).

The insurance industry uses predictive analytics for setting policy premiums. Many automotive insurance companies can accurately predict the likelihood a driver will be in an accident or have their car stolen. The insurance companies gather driver's behavior data from databases, transmissions from installed boxes in a car, or from an app on the driver's smartphone. Insurance companies also use predictive analytics in profiling fraudulent claims. When a claim is submitted, certain variables within the claim are compared to past fraudulent claims. When certain variables pair up with the fraudulent claim, the submitted claim can be flagged for further investigation. Insurance companies applying predictive analysis techniques on data collected from social media, credit agencies, and other open sources of demographic data can profile individuals who are likely to commit fraud (Marr 2015).

In an effort to improve estimated time of arrival (ETA), a major US airline using publicly available data about the weather, flight schedules, and other information along with the company's proprietary data was able to use predictive analysis to answer two questions—"What happened all the previous times when a plane approached this airport under these conditions?" and "When did it actually land?" After implementing this analysis, the airline was able to eliminate the difference between the estimated time of arrive and the actual time of arrival resulting in significant savings to the airline (McAfee and Brynjolfsson 2012).

Throughout the Covid-19 pandemic, healthcare providers have used predictive analytics to develop models to better understand patient risks for contracting the disease and to predict the severity of the disease based upon the characteristics of the patient. Researchers at Mount Sinai created a model based upon age, minimum oxygen saturation, and type of patient encounter to try to minimize the severity of the disease within patients they treated (Kent 2020). Other researchers built models to predict who would likely test positive, how the disease would spread, and what other factors could predict the risk and/or severity for contracting the disease. Analytics has other applications within the healthcare industry, especially as both the data and the predictive algorithms become more sophisticated. For example, the Internet of Things (IoT) along with big data have made it possible to monitor patients' heart rates, oxygen levels, sleep rates, glucose levels, and blood pressure to predict when patients are more likely to experience serious medical issues to try to prevent or treat them earlier. Recommended actions can be provided that engage patients more fully in their healthcare decisions.

The healthcare industry also uses predictive analytics in a variety of ways including predicting hospital readmissions and managed care and even identifying potential opioid abuse. BlueCross BlueShield of Tennessee (BCBS) partnered with Fuzzy Logix to identify characteristics of individuals at risk of abuse. The data input into the predictive analytics model included years of pharmacy data, claims data from both BCBS's customers, and others. The results provided insightful behaviors that could predict whether an individual could be at risk of developing an opioid abuse problem. Some of these behaviors included frequent use of different doctors and pharmacies. A total of 742 variables were used, and the predictions that can be made are approximately 85% accurate. Hopefully, armed with this information preventive campaigns can be launched and doctors educated on advising their patients who may be at risk (Marr 2017).

Both internal and external auditing practices are changing as well. Historically, audits used small samples at a point in time (e.g., annual); however there is a surge in applying predictive analytics in audit engagements. Auditors can now review the complete data population and engage in continuous auditing. Applying predictive analytics increases the efficiency and improves the audit. In internal audit, predictive analytics can assist in monitoring numerous regulations, practices, and procedures as well as detect risk and fraud (Syvaniemi 2015). In external audit, predictive analytics can assist auditors in discovering business risks and fraud detection and evaluating going concern of an entity (Earley 2015).

In 2011 the Internal Revenue Service (IRS) created a new division called the Office of Compliance Analytics. The purpose of the new division is to create an advanced analytics program that will use predictive analytics algorithms to reduce fraud. The IRS collects commercial and public data, including information from Facebook, Instagram, and Twitter. Algorithms are run against their collected information combined with the IRS' proprietary databases to identify potential noncompliant taxpayers (Houser and Sanders 2017).

Police departments are using predictive analytics (referred to as predictive policing) to optimize crime prevention strategies. Predictive policing is proactive and can help answer questions such as where is violent gun crime likely to occur? Where is a serial burglar likely to commit his next crime? Where are criminals located? Data sources include past crime data, population density, the presence of expensive cars, the presence of tourists, payday schedules, weather and seasonal patterns, traffic patterns, moon phases, time of day, weekend or weekday, escape routes (e.g., bridges, tunnels, dense foliage), and social networks (e.g., family, friends, affiliations). Applying predictive analytics helps police departments create crime maps which forecast criminal hot spots, areas where crime is likely to occur. Then the police departments can allocate resources and increased focus in these areas. Predictive analytics is also employed in tracking criminals after the crime has been committed (Bacchner 2013).

The city of Chicago used predictive analytics to solve rat infestation problems. By examining the city's 311 data, a team from the Department of Innovation and Technology observes a correlation between 311 (non-emergency municipal calls) call types and rat infestation problems. The team developed a predictive analytics model that provided specific areas that targeted where intervention (rat baiting) was required. When the Department of Streets and Sanitation (DSS) utilized the information, they uncovered the largest infestation they had ever seen (Goldsmith 2016).

From the examples above, we can see how predictive analytics has penetrated into virtually every industry and every functional organization within each industry. The examples illustrate a variety of ways in which organizations have achieved real value through their use of predictive analytics.

1.2   Analytics Landscape

The analytics landscape includes terms like business analytics, big data analytics, data mining, data visualization, and big data. Gartner, a leading information technology research and advisory company, defines business analytics as follows:

"Business Analytics is comprised of solutions used to build analysis models and simulations to create scenarios, understand realities and predict future states. Business analytics includes data mining, predictive analytics, applied analytics and statistics (Gartner n.d.-a)." Today, business analytics is essential as it can improve an organization's competitive edge, thus enhancing revenues, profitability, and shareholder return.

IBM defines *big data analytics* as "the use of advanced analytic techniques against very large, diverse data sets that include structured, semi-structured and unstructured data, from different sources, and in different sizes from terabytes to zettabytes (IBM n.d.)."

Data mining is applying advanced statistical techniques to discover significant data patterns, trends, and correlations in large data sets with the overarching goal of gathering insights previously unknown about the data sets and transforming the data for future use.

Figure 1.1 illustrates the relationships between big data, data mining, and analytics. These three terms overlap and collectively are considered business analytics. Basically, business analytics is the use of advanced analytic techniques to discover meaningful insights from large, complex data sets in an opportunistic timeframe.

Fig. 1.1 Business
analytics

1.2.1   Big Data

When individuals hear the term big data, typically they think of a large amount of data and that it is just size that matters. Most individuals recognize megabytes, giga-bytes, or even terabytes. Today, large companies are storing transactions in pet-abytes. In a SAS white paper Big Data Meets Big Data Analytics, SAS states the following:

- "Wal-Mart handles more than a million customer transactions each hour and imports those into databases estimated to contain more than 2.5 petabytes of data.
- Radio frequency identification (RFID) systems used by retailers and others can generate 100–1000 times the data of conventional bar code systems.
- Facebook handles more than 250 million photo uploads and the interactions of 800 million active users with more than 900 million objects (pages, groups, etc.) each day.
- More than five billion people are calling, texting, tweeting and browsing on mobile phones worldwide." (Troester n.d.)

UPS has been collecting and working with big data for a long time and has over 16 petabytes of data stored. They collect data from their vehicles including but not limited to engine performance and condition, speed, number of stops, mileage, and miles per gallon. They collect GPS data, delivery, and customer services data and address points and routes traveled. Their drivers have handheld devices that also collect data (van Rijmenjam 2018).

How big is a petabyte? Let's begin first with the smallest unit of storage, a bit. A bit represents a single binary digit (a value of 1 or 0). The smallest grouping of bits that has meaning is a byte. Data size is measured in increments based upon increases in the number of bytes. To understand how big is a petabyte, Table 1.1 provides a brief description of data sizes.

Table 1.1 Large data sizes

Unit	Approximate size (decimal)	Examples
Bytes (B)	8 bits	One byte = one character; 10 bytes = one word
Kilobyte (KB)	1000 bytes	Two KBs = a typewritten page
Megabyte (MB)	1000,000 bytes	One MB = a small novel; five MBs = complete work of Shakespeare
Gigabyte (GB)	1000 megabytes	16 h of continuous music
Terabyte (TB)	1000 gigabytes	130,000 digital photos
Petabyte (PB)	1000 terabytes	Two PBs = all US academic research libraries
Exabyte (EB)	1000 petabytes	Five EBs = estimate of all words ever spoken by human beings
Zettabyte (ZB)	1000 exabytes	36,000 years of high-definition video
Yottabyte (YB)	1000 zettabytes	

Data scientists are now looking toward yottabytes of storage. Big data is not just about the size of the data; other factors come into consideration. While there is no universal definition of big data, Gartner's definition is widely used. "*Big data* is high-volume, high-velocity and/or high-variety information assets that demand cost-effective, innovative forms of information processing that enable enhanced insight, decision making, and process automation" (Gartner n.d.-b). Figure 1.2 illustrates the three defining properties (3Vs) of big data. High volume refers to the magnitude of the data. There is no set standard of what process and storage size constitute big data. Big data is relative. Some organizations think terabytes are big data, while others think of petabytes as big data. What does the future look like? Well, terabytes and petabytes may be considered small data sets in the future as storage capacities increase. Additionally with the increasing use of the IoT (Internet of Things), physical objects other than computers that contain software and sensors are wired or have wireless connectivity. Examples include smart cars, smartphones, smart thermostats, fitness watches and equipment, and appliances. An enormous amount of data is being collected and is available for analysis.

High velocity is of course speed—the speed of creating data, the speed of storing data, and the speed of analyzing data to make decisions rapidly and opportunistically. Historically organizations used batch processing. Large chunks of data would be input into mainframe computers and take a significant amount of time to process and eventually receive the results. This is useful only if timeliness is not an issue for decision-making. Further, computationally resource-intensive algorithms were not possible or took so long to execute few organizations that would even attempt to use them. However, in today's highly competitive markets, real-time (streaming) processing is necessary. Real-time processing matches the speed of data input with the speed of processing and the speed of information generation and enables companies to have real-time decision-making power. This creates value and a competitive advantage.

Fig. 1.2 Big data

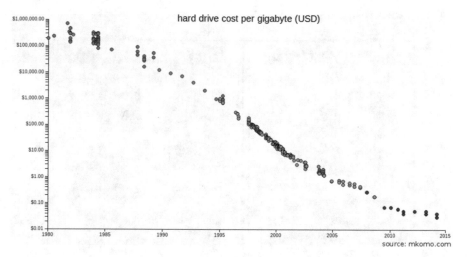

Fig. 1.3 Hard drive cost per gigabyte (Komorowski 2014)

High variety points to the many different types of data. Historically, data was structured and stored in relational databases. Unstructured data, while available, was not easily accessible. Now, data is structured, semi-structured, and unstructured. The cost of data storage has significantly decreased over the past 30 years (see Fig. 1.3).

Structured data is highly organized data, usually in columns and rows, and can easily be searched and processed by straightforward data mining techniques. Think of spreadsheets and relational databases. Traditional database systems include Oracle, SQL Server, and DB2. Unstructured data is unorganized and cannot easily be identified by rows and columns. Digital images, e-mail, PDF files, videos, audios, tweets, Google searches, and Facebook likes are examples of unstructured data. Unstructured data presents new challenges to store and analyze data. This is why new products such as Hadoop have emerged. Semi-structured data lies in between structured and unstructured data. It does not adhere to a formal data structure yet

Fig. 1.4 Types of data

Fig. 1.5 Examples of structured and unstructured

does contain tags and other markers to organize the data. XML (Extensible Markup Language) is an example of semi-structured data (Figs. 1.4 and 1.5).

In summary, the terms business analytics, data mining, and big data are all related and contain overlapping perceptions of the same ideas and concepts in their definitions. Additionally, the descriptions include analysis of the data in their definitions. Organizations have amassed a vast amount of data that they want to make use of. The next sections will discuss the analytics used in business analytics.

1.3   Analytics

Analytics are applied to obtain useful information from the data. The fundamental goal of business analytics is to make better decisions based on the data. In this section, the two main types of analytics will be discussed, descriptive analytics and predictive analytics. Additionally, diagnostic and prescriptive analytics will be briefly mentioned. Descriptive analytics is usually the starting point of analysis performed on the data.

Descriptive analytics is a very basic form of analytics. It summarizes the raw data and describes the past. In other words, "What has happened?" Descriptive analytics is useful in providing information about past behaviors, patterns, or trends in the data. Descriptive analytics also aids in the preparation of data for future use in prescriptive analytics. It includes information like sums, averages, and percent changes. Business examples include the sum of sales by store, total profit by product or distribution channels, average inventory, number of complaints resolved in the past quarter, classification by customer, or the average amount spent per customer. Walmart uses descriptive analytics to uncover patterns in sales, determine what customers buy online versus in the store, and see what is trending on Twitter. Chapter 3 discusses descriptive analytics in detail.

While descriptive analytics describes what has happened, diagnostic analytics seeks to answer "Why did it happen?" Diagnostic analytics attempts to get at the root cause of some anomaly or occurrence; for example, why did sales increase in the northeast when there was no special marketing campaign? Some techniques used include drilling down on the data, correlations, and looking outside of the data set for other causes.

1.3.1   Predictive Analytics

Simply put, predictive analytics uses historical data to predict future events. The central question for predictive analytics is "What will happen?" Predictive analytics uses advanced statistics and other machine learning techniques. It is essential that historical data be representative of future trends for predictive analytics to be effective. Predictive analytics techniques can be used to predict a value. How long can this airplane engine run before requiring maintenance or to estimate a probability? How likely is it that a customer will default on a mortgage loan? Predictive analytical techniques can also be used to pick a category. What brand of sneakers will the customer buy? Nike, New Balance, or Adidas? Predictive analytics uses data-driven algorithms to generate models. Algorithms are step-by-step processes for solving problems. Algorithms take the data through a sequence of steps to calculate predictive results. Effectively, predictive analytics algorithms automate the process of discovering insights (e.g., patterns, trends) in the data.

Predictive analytics algorithms are split into *supervised learning algorithms* and *unsupervised learning algorithms*. Supervised learning algorithms involve an iterative process of learning from the training (historical) data set. The training data set contains labels and has the correct information (i.e., the information the model is learning to predict). The prediction output is referred to as the target variable. As the predictive model is trained and reaches an optimal point, the model is now ready to produce predictive output. For example, a non-profit organization is interested in identifying the most-likely, potential donors. A historical data set containing past donors and their characteristics and demographics could be used. Unsupervised learning algorithms involve data sets that have no target variable. The data is not classified or labeled. In unsupervised learning a prediction cannot be made; therefore, the data is analyzed, and the results are grouped into clusters (e.g., group customers by their browsing and purchase behavior). Throughout this book supervised learning algorithms will be used.

Predictive analytics modeling techniques fall into two major categories: regression techniques and machine learning techniques.

1.4 Regression Analysis

Regression analysis is a widely used technique in predictive analytics. Regression analysis is a technique for measuring relationships between two or more variables and can be used to predict actual outcomes. The value that you want to predict the outcome is the dependent variable ("The effect of" and typically shown as Y in the regression equation. Also known as the target variable in predictive analytics). Independent variables, predictor variables, are inputs into the regression equation that are assumed to have a direct effect on the target variable ("Cause of" and typically shown as an X in the regression equation). Regression techniques provide a mathematical equation that describes the relationship between the target (predictor) variable and the other independent variables.

Two types of regression techniques will be demonstrated in Chap. 4 of this book. They are linear regression and logistic regression. Linear regression is used when the target variable is continuous and can have any one of an infinite number of values. Linear regression uses a straight line and therefore requires a linear relationship between the target variable and independent variable. Logistic regression is used when the target variable is categorical or discrete. The target variable in logistic regression is binary. In other words, the output is either of two outcomes, e.g., yes or no. Figure 1.6 shows a graphical comparison between the two regression methods. Notice that the linear regression equation is shown in the form of a straight line that best fits the independent variables (x) with the target variable (y). The logistic regression equation when graphed looks like an S.

Fig. 1.6 Linear regression
graph and logistic
regression graph

1.5   Machine Learning Techniques

Machine learning, a branch of artificial intelligence, uses computation algorithms to automatically *learn* insights from the data and make better decisions in the future with minimal intervention. Regression techniques may also be in machine learning techniques, e.g., neural networks. An *artificial neural network (ANN)* or *neural network* is a system, including algorithms and hardware, that strives to imitate the activity of neurons in the brain. A key element of the neural network is the ability to learn from the data. The neural network is composed of multiple nodes and is typically organized in layers. The layers are composed of interconnected nodes. The

Fig. 1.7 Neural network

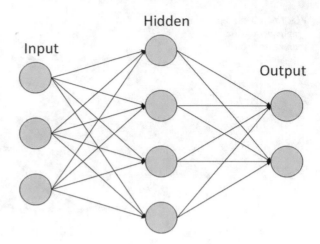

nodes can take input data and perform algorithms on the data and then pass the results (activation or node value) to intermediary or hidden layers. The hidden layers then link to an output layer reflecting the results. Figure 1.7 is an example of a simple neural network. Chapter 6 will examine neural networks.

Decision trees are very popular in predictive modeling for several reasons—they are easy to understand, are easy to build, can handle both nominal and continuous variables, can handle missing data automatically, and can handle a wide variety of data sets without applying transformations to the input. Decision trees are composed of nodes and branches. The nodes contain questions (premises), and the response to the questions determines the path to the next node (conclusion). Branches link the nodes reflecting the path along the decision tree. The initial question on the decision tree is referred to as the root node. Figure 1.8 provides an example of a decision tree. Chapter 5 will examine decision tree models in detail.

The premise of predictive analytics is that it is possible to model data, and underlying this premise is the notion that a cause-and-effect relationship exists between the data boundaries. That is, as some data boundaries change (cause), there will be a corresponding change to the other data boundaries (effect).

In addition to having knowledge of the statistical algorithms and machine learning tools, three other components are necessary to create value in predictive analytics (Fig. 1.9).

The first component is strong business knowledge. Typically, a predictive analytics project will be team-based including members with in-depth knowledge of the organizations' industry and strategies as well as data science and analytics experts. The organization experts will have a sense of the characteristics and relationships that may exist between the variables in the data. The data scientists and analysts possess in-depth analysis techniques to create models to gain meaningful insights, including behavior patterns and trends from the data. The second component requires that a complex business problem exists that necessitates advanced analytic techniques to solve. Third is access to big data whether from within the

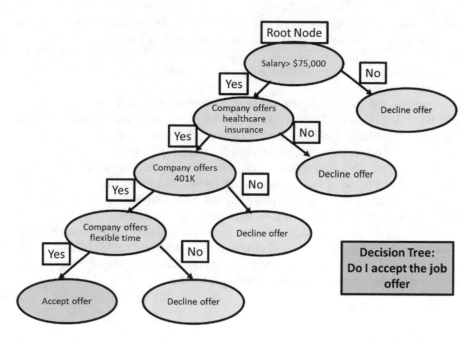

Fig. 1.8 Decision tree

Fig. 1.9 Predicative
analytics value creation

organization, outside of the organization, or both. Technological advances in data storage, data processing, and data mining have made it possible to finally take advantage of large data sets to discover opportunities such as emerging trends, markets, and customers. When predictive analytics is applied, the result is value added to the organization.

Applying predictive analytics to big data enables organizations to make better and faster decisions. Predictive analytics can help organizations mitigate their risk, maximize their profits, optimize their operations, and gain a competitive advantage.

The next frontier takes predictive analytics models further by recommending a course of action. This occurs once an estimate of what will happen in the future is modeled, thus informing the decision-maker on how to react in the best way possible given the prediction. Some have argued that informing the decision-maker is a separate category of analytics referred to as prescriptive analytics.

Data itself does not have value. Rather it is what you do with the data that adds value. Again, organizations use predictive analytics to solve complex business problems, discover new opportunities, and gain competitive advantages. Examples of common business problems are presented in Table 1.2.

Table 1.2 Common business problems

Detection of fraud	Banking and financial service companies need to know how to detect and reduce fraudulent claims
	Audit firms need to know how to identify fraudulent business transactions
	Health and property-casualty insurance companies need to know which claims are fraudulent
Improve operational efficiency	Manufacture and retail companies need to know how to create better inventory forecasts
	Companies need to know how to improve resource management
	Hotels need to know how to maximize occupancy rates and what room rate to charge per night
	Airline companies need to optimize the timing of when aircraft engines require maintenance
Optimization of market campaigns/customer insights	Retail companies need to determine customer responses, purchase behavior, cross-sell opportunities, and effectiveness of promotional campaigns
	Customer loyalty programs target customers at the right time to maximize their purchases
Reduce risk	Banking and financial service companies currently rely on credit scores to determine a customer's creditworthiness
	Health insurance companies need to know which individuals are more at risk for chronic diseases
Enhance innovation	Many companies need to constantly bring new products to the market
	Pharmaceutical companies must continually innovate for new drug discovery

1.6   Predictive Analytics Model

Figure 1.10 provides a nine-step model for using predictive analytics. The first step in the model is to identify the business problem. Business-specific knowledge and problem identification are critical to developing a successful predictive analytics model. Additionally, knowledge of the data storage and infrastructure, as well as proficiency in predictive modeling, is required. Frequently this will require a team of experts in each of these areas. The ability to solve these business problems rapidly may lead to higher revenues or lower expenses resulting in a competitive advantage.

Step 2 in the predictive analytics model is to define the hypotheses. The hypotheses are developed from the business problem. The purpose of the hypotheses is to narrow down the business problem and make predictions about the relationships between two or more data variables. The hypotheses should clearly state what is going to be analyzed, i.e., predicted. Table 1.3 provides some examples of transitioning from a business problem to a hypothesis.

After the business problem and hypotheses are identified, step 3 is to collect the data. The data can be structured (e.g., tabular data found in spreadsheets or relational databases), semi-structured (e.g., Extensible Markup Language (XML)), or unstructured data (e.g., text, images, audio, and video). Data can be sourced from within the organization (e.g., transactional databases or data warehouses), data can be purchased (e.g., Acxiom, Nielsen, Experian, Equifax), or data can be obtained

Fig. 1.10 Predictive analytics model

1. Define Business Problem
2. Define Hypotheses
3. Collect Data
4. Analyze Data
5. Develop Predictive Model
6. Optimize Model
7. Determine Best Fit
8. Utilize Model/Score New Data
9. Monitor Model

Table 1.3 Business problem and hypothesis

Business problem	Hypothesis example
How to detect and reduce fraudulent claims	Gender, education, marital status, and income are indicators of individuals who would commit insurance fraud
Need to increase sales	Amount of time spent on company website determines purchase Facebook advertising generates more sales compared to LinkedIn advertising The word "call" in a marketing campaign results in increased leads
Filling a prescription takes too long; need to know how much staff to have available	The day social security checks are released are the busiest days The hours between 9 a.m. and noon are the busiest
Need to know what individuals will purchase the next-generation phone for marketing campaign to gain potential buyers	Age, prior product use, income, education, and location will indicate individuals who will purchase new phone

from free resources (e.g., data.gov). More insights can be obtained by merging different data sets from different sources.

After collecting the data, step 4 requires the data to be analyzed. This step in the process is likely to be the most time-consuming. It is a critical analytical step. Data quality is key. If your data is dirty, then as the saying goes *garbage in, garbage out*. Often the data is collected from multiple sources and contains different information. The data sources need to be assimilated and synchronized to ensure consistent format and include the appropriate data for analyses. Next transformation methods are applied to the raw data. The raw data may contain missing values, redundant variables, outliers, and erroneous data. If the data has missing variables, a decision should be made whether to omit the record or apply some other method to replace the value. Outliers can be viewed from frequency distribution tables, histograms, box plots, or scatter plots, and then decisions need to be made on whether to include the outliers. Sometimes analysis of outliers can be key to the problem; other times they merely add "noise" to the data. If variables have wide ranges, then the data should be categorized. Duplicate records and duplicate variables should be removed. Methods for handling these situations are discussed in Chaps. 2 and 3. In Chap. 8 cluster analysis is presented to analyze data to look for relationships and associations that may be previously unknown. Since there are often large volumes of text data that accompany quantitative data, Chap. 9 describes how text mining can be used to analyze qualitative data.

Step 5 in the process is to develop the predictive model. This step in the process typically requires building and testing multiple models including regression techniques and machine learning techniques (e.g., neural networks). The goal is to create a useful model; therefore it is important to ensure you have the right (independent) variables and correct weights. This will be covered in more detail in Chaps. 2, 3, 4, 5, 6, and 7.

Step 6 in the process is to optimize the model. Optimizing the model aims to improve the model's predictability. In this step parameters of the model are modified based upon the data. This can be an iterative process. For example, with a neural network, if the network is too simple, then the number of hidden layers may need to be increased. Or if the data does not have complex relationships, adding direct connections may improve the model. Optimizing the model will be discussed along with the "Big 3" techniques—linear regression, neural networks, and decision trees in Chaps. 4, 5, and 6.

Step 7 is to determine the best-fit model. Typically, in predictive modeling, multiple techniques such as regression, decision trees, or neural networks (the Big 3) are used. Within each technique, multiple models may be used. For example, multiple neural networks or multiple regression techniques are utilized. Based upon the selection statistic, i.e., criteria, the model with the most accurate predictive capability is chosen. The process to determine the best-fit model is discussed in Chap. 7.

Step 8 is to utilize the model, referred to as scoring. Up through step 7, historical data has been used and the answer is known. The purpose has been to build and train the models. Simply put, the answer needs to be known to determine how close the calculated answer is. At the point where the calculated answer is close enough, then the model can be used to apply against a current business data set for the purpose of making predictions about the current data and ultimately make business decisions based on the predictions. Utilization and scoring the model are covered in Chap. 7.

Step 9 is to monitor the model. Consistent with any other information system, a process is needed to continually review and adjust the models for new information as needed. The results of the model should also be continuously monitored and validated for accuracy.

1.7 Opportunities in Analytics

Similar to how the personal computer transformed the business world over 30 years ago, data science and analytical skills are creating a new disruptive force in the workplace. Currently, data scientists and individuals with analytic skills are in hot demand. McKinsey Global Institute (2011) predicted a 140,000 to 190,000 gap in the supply of analytic talent. Further, they forecast in the United States a demand for 1.5 million additional managers and analysts who have the skills to ask the right questions and interpret the results. O*Net (2018), a comprehensive source of occupational information sponsored by the US Department of Labor reported that there were 287,000 employees in the United States with a job title business intelligence analyst with the demand for an additional 22,400 projected future job openings. The US Bureau of Labor and Statistics lists data scientists, statisticians, and operations analysts among its hottest growing job opportunities. A Gallup poll, administered for the Business-Higher Education Forum (PWC and BHEF 2017), exposed a significant disparity between student supply with data science and analytical skills and employers' expectations. The findings showed that by 2021, 69% of employers expect candidates with data science and analytics skills to get preference for jobs in their organizations. However, only 23% of college and university leaders say their graduates will have those skills (Fig. 1.11).

There are many different career paths to choose from. For example, a data analyst career path may include examining large data sets, identifying patterns and trends in the data, and providing insights that influence better decision-making. This includes developing charts and visualizations to communicate the insights, whereas a data scientist designs and develops new processes for modeling including creating algorithms and predictive models. However, increasingly organizations are looking for accounting, finance, marketing, healthcare, human resources, supply chain, and logistics professionals, to complement their disciplinary skill set with data science and analytics skills. In fact, a Gallup poll administered for Business-Higher Education Forum (PWC and BHEF 2017) revealed that by 2020:

- 59% of finance and accounting managers
- 51% of marketing and sales managers
- 49% of executive leaders
- 40% supply chain and logistics managers

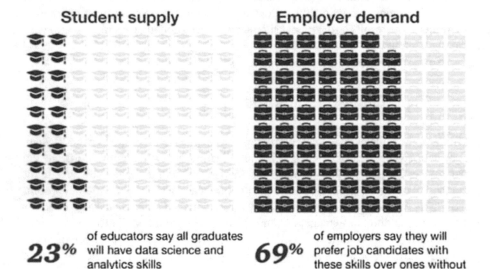

Data science and analytics skills, by 2021
How will employers fill the talent pipeline?

Student supply

Employer demand

23% of educators say all graduates will have data science and analytics skills

69% of employers say they will prefer job candidates with these skills over ones without

Base: Higher education: 127; Business: 63
Source: Gallup and BHEF, Data Science and Analytics Higher Education Survey (December 2016).

Fig. 1.11 Data science and analytics skills (Gallup and BHEF, Data Science and Analytics Higher Education Survey (Dec 2016))

- 30% of human resources managers
- 48% of operations managers

be required to possess data science and analytics skills. The technical skills can be acquired through various university programs; however to be a good analyst requires a specific type of critical thinking mindset. What is a *critical thinking mindset*? Individuals with a critical thinking mindset tend to be inquisitive, curious, and always asking "why." A critical thinker demonstrates an optimistic attitude and the motivation to solve complex problems. In fact, they get excited when presented by the problem. A critical thinker is willing to take risks and challenge the status-quo, is creative, and suggests new ideas. A critical thinker does not give up, remains optimistic, and pushes through. The critical thinker is tenacious. The critical thinking mindset also includes being data and business savvy, including understanding the business problem the organization is trying to solve, asking the right questions, understanding how to create the appropriate data analysis, demonstrating an ability to be an effective team player, and possessing excellent presentation and communication skills.

Individuals with discipline-specific expertise coupled with strong understanding of data analytics and ability to leverage data to generate better decisions will create organization value and competitiveness. These individuals will be invaluable to organizations in the future.

1.8 Introduction to the Automobile Insurance Claim Fraud Example

The cost of fraud to the property-casualty insurance industry is tremendous and continues to increase globally. According to the National Insurance Crime Bureau, fraud adds 5–10% to the average insurance premium in the United States. As a result, every policyholder pays for fraudulent claims. Fraud is a recurrent problem that has existed since the inception of the property-casualty insurance industry. Insurance companies must be vigilant in identifying and managing fraudulent claims. Claim fraud analysis is one of the key analytics for property-casualty insurers, and most insurers have a dedicated Special Investigative Unit (SIU) to investigate and resolve potentially fraudulent claims (Saporito 2015). According to the Insurance Information Institute (2018), 42 states and the District of Columbia have set up fraud bureaus for the purpose of reporting potentially fraudulent claims. In some cases, they are set up by line of business resulting in multiple bureaus. Although fraud continues to grow across all lines of business in the insurance industry, healthcare, workers compensation, and automobile insurance have been the three most prevalent lines of business to experience fraudulent claims. Insurance fraud continues to be a significant challenge for the industry, regulatory authorities, and the public worldwide. Predictive analytics utilizing data-driven fraud detection offers the possibility of determining patterns that uncover new potentially fraudulent claims which can then be investigated. We now have the capability of analyzing a massive amount of historical data to predict the potential for fraud in current and future claims. This can provide both a cost efficiency (through the reduction in claim payments) and workload efficiency (Baesens et al. 2015).

Throughout the remainder of this book, an auto insurance claim data set will be used to create predictive models to evaluate potentially fraudulent claims. The data set comprises historical claim data with variables considered to be significant in the identification of potentially fraudulent claims. The book will walk through the predictive analytics model applying the auto insurance claim data set to the "Big 3," as well as several other techniques to predict potentially fraudulent claims. SAS Enterprise Miner™ will be the software application to apply these techniques. See Appendix A for the detailed description of the data. Figure 1.12 highlights the first three steps in the predictive analytics model.

Step 1 in the model is to determine the business problem. The business problem in this example is to develop a predictive model to detect and mitigate potentially fraudulent automobile insurance claims.

Step 2 in the model is to narrow down the business problem and develop the hypotheses. Recall that the hypotheses will make predictions about the relationships between two or more data variables. For the business problem, the hypotheses collectively will be:

> State code, claim amount, education, employment status, gender, income, marital status, vehicle class, and vehicle size will predict a fraudulent claim.

These are the independent (predictor, input) variables to be used in the model. The independent variables are assumed to have a direct effect on the dependent variable. For this example, the target variable is a fraudulent claim indicator. Please see Appendix A for a list of all the variables and their definitions provided in the data set.

Fig. 1.12 Predictive analytics model – Steps 1, 2, 3

Step 3 is to collect the data. The auto insurance claim data is a structured data set that will be used throughout the book in creating various models to predict fraudulent claims.

The remainder of this book explains in further detail the steps in the predictive analytics model and various techniques to build the "Big 3" predictive models using the auto insurance claim data set. Chapter 2 reviews data preparation and analyses. Chapter 3 examines descriptive statistics. Chapter 4 presents linear and logistic regression modeling techniques. Chapter 5 discusses decision trees. Chapter 6 describes neural networks and how to optimize them. Chapter 7 presents other predictive modeling techniques, model comparisons, and scoring and will cover the predictive analytics model steps 7, 8, and 9. Chapter 8 describes how cluster analysis is used to analyze data to look for unknown relationships and associations. Chapter 9 describes qualitative data can be analyzed through the use of text mining.

1.9 Chapter Summary

New innovations in computer processing speeds, cheaper storage, and massively parallel processing (MPP) architectures have facilitated organizations applying advanced analytical techniques against enormous data sets. Big data combined with the ability to analyze the data in real time or near real time positions companies to capitalize on the insights that can be discovered in their vast amount of historical data. Terms to describe this phenomenon such as big data, business analytics, and big data analytics have emerged. All these terms ultimately aspire to glean insights from the data. Descriptive analytics analyzes the data to see what has happened in the past. Organizations use descriptive analytics to uncover past behavior, trends, and patterns. Diagnostic analytics looks at the historical data and attempts to explain why the trends and patterns occurred. The real value for organizations though rests in the application of predictive analytics. Predictive analytics applies advanced analytic techniques to historical data to predict future events, "What will happen?" Predictive analytics uncovers behaviors, patterns, and trends and produces a prediction or probability of a likely outcome.

The "Big 3" predictive analytics techniques include regression, decision trees, and neural networks in creating algorithms and predictive models. Two very popular regression techniques include linear and logistic regression. Linear regression is utilized when the target variable is a continuous and target variable, and independent variables have a linear relationship. Logistic regression is used when the target variable is binary and categorical.

Neural networks are a machine learning technique that uses advanced algorithms to automatically "learn" insights from historical data. Neural networks mimic the activity of neurons in the human brain.

Decision trees are popular due to their ease in construction and understanding. They require less intensive data preparation. A decision tree uses nodes, branches, and leaves to arrive at and communicate a predictive result.

The application of predictive analytics techniques can create a tremendous value for organizations. Today, many organizations apply significant resources and gain added value including higher profits, efficient operations, fraud prevention and detection, effective marketing campaigns, and other competitive advantages. To reap the tremendous benefits from predictive analytics, organizations must also have a complex business problem, vast amounts and a variety of data coupled with the capability to process and analyze swiftly, and business knowledge expertise. These three components, complex problem, big data, and business knowledge, when exploited together provide impressive value to companies. This chapter examined some of the leading companies that have gained competitive advantages from incorporating predictive analytics into their business including Amazon, Target, UPS, and Walmart. Most industries are ramping up their predictive analytics skill set to remain competitive. Government agencies and non-for-profits are also utilizing predictive analytics. It will become an environment of "adapt or die."

A nine-step predictive analytics model was introduced that outlined an effective step-by-step process to create predictive models. The steps were the following: Define the business problem, narrow the business problem down to the hypotheses, collect the data, analyze (clean, prepare) the data, develop a predictive model, optimize the model, determine the best-fit model, score the data, and finally monitor the model.

Career opportunities are vast in data analytic positions. Every major business discipline is seeking analytics skills. Though the technical skills can be learned, a critical thinking mindset is required to be an effective data scientist or data analyst. A critical thinking mindset includes being inquisitive, optimistic, motivated, risk-taker, creative, and tenacious. A critical thinking mindset also includes the ability to work well in teams and excellent communication and presentation skills.

Lastly, throughout this book the nine-step predictive analytics model will be applied using an auto insurance claim data set. Using the "Big 3" analytic techniques, predictive models will be built with SAS Enterprise Miner™. Using historical claim data, models will be built to predict whether future claims could be fraudulent.

Incorporating predictive analytics within an organization with access to big data identifies a complex problem and possesses business expertise that can facilitate better decision-making creating value for an organization.

Discussion Questions
1. Describe big analytics, big data, and predictive analytics. What do they have in common?
2. Describe an example of an organization that uses predictive analytics and how it has added value to the company.
3. Discuss the nine steps of the predictive analytics model.
4. Why is a critical thinking mindset important in a career in analytics?
5. Discuss at least three different possible career opportunities enabled by predictive analytics.

References

Bacchner J (2013) Predictive policing preventing crimes with data and analytics. White paper from the IBM Center for the business of government. Available at http://businessofgovernment.org/sites/default/files/Predictive%20Policing.pdf. Accessed 15 June 2021

Baesens B, Van Vlasselaer V, Verbeke W (2015) Fraud analytics: using descriptive, predictive, and social network techniques: a guide to data science for fraud detection. Wiley & Sons, Hoboken

Carr D (2013) Giving viewers what they want. http://www.nytimes.com/2013/02/25/business/media/for-house-of-cards-using-big-data-to-guarantee-its-popularity.html. Accessed 15 June 2021

Earley C (2015) Data analytics in auditing: opportunities and challenges. Bus Horiz 58:493–500

Gartner (n.d.-a). https://www.gartner.com/it-glossary/business-analytics/. Accessed 15 June 2021

Gartner (n.d.-b). https://www.gartner.com/it-glossary/big-data. Accessed 15 June 2021

Goldsmith S (2016) The power of data and predictive analytics. Government Technology. http://www.govtech.com/data/The-Power-of-Data-and-Predictive-Analytics.html. Accessed 15 June 2021

Hays C (2004) What Wal-Mart knows about customers' habits. New York Times. Nov 14, 2004. https://www.nytimes.com/2004/11/14/business/yourmoney/what-walmart-knows-about-customers-habits.html. Accessed 15 June 2021

Houser K, Sanders D (2017) Vanderbilt. Journal of Entertainment and Technology Law, 19(4). Available via http://www.jetlaw.org/wp-content/uploads/2017/04/Houser-Sanders_Final.pdf. Accessed 15 June 2021

IBM (n.d.) What is big data analytics? https://www.ibm.com/analytics/hadoop/big-data-analytics. Accessed 15 June 2021

Insurance Information Institute (2018) Background On: Insurance Fraud. https://www.iii.org/article/background-on-insurance-fraud. Accessed Jan 1, 2018

Kent J (2020) 3 ways healthcare is using predictive analytics to combat COVID-19, Health IT Analytics. https://healthitanalytics.com/news/3-ways-healthcare-is-using-predictive-analytics-to-combat-covid-19. Accessed 15 June 2021

Komorowski M (2014) A history of history of storage cost (update). http://www.mkomo.com/cost-per-gigabyte-update. Accessed 9 Sept 2018

Kopalle P (2014) Why Amazon's anticipatory shipping is pure genius. Forbes online January 28, 2014. https://www.forbes.com/sites/onmarketing/2014/01/28/why-amazons-anticipatory-shipping-is-pure-genius/#3b3f13704605. Accessed 9 Sept 2018

Marr B (2015) How big data is changing the insurance industry forever. Forbes online December 16, 2015. Retrieved at https://www.forbes.com/sites/bernardmarr/2015/12/16/how-big-data-is-changing-the-insurance-industry-forever/#28c1225d289b

Marr B (2017) How Big Data helps to tackle the number 1 cause of accidental death in the U.S. Forbes online January 17, 2017 https://www.forbes.com/sites/bernardmarr/2017/01/16/how-big-data-helps-to-tackle-the-no-1-cause-of-accidental-death-in-the-u-s/#76e20ad639ca. Accessed 9 Sept 2018

McAfee A, Brynjolfsson (2012) Big data: the management revolution. Harvard Business Review. October 2012

McKinsey Global Institute (2011) Big data: the next frontier for innovation, competition, and productivity. https://www.mckinsey.com/business-functions/digital-mckinsey/our-insights/big-data-the-next-frontier-for-innovation. Accessed 9 Sept 2018

O*Net (2018). https://www.onetonline.org/link/summary/15-1199.08?redir=15-1099. Accessed 9 Sept 2018

Power B (2017) How Harley-Davidson used artificial intelligence to increase New York sales leads by 2,930%. Harvard Business Review. May 30, 2017. https://hbr.org/2017/05/how-harley-davidson-used-predictive-analytics-to-increase-new-york-sales-leads-by-2930. Accessed 9 Sept 2018

Price-Waterhouse Coopers & Business-Higher Education Forum (2017) Investing in Americas data science and analytics talent. Retrieved at pwc.com/us/dsa-skills. Accessed 15 June 2021

Reuters (2018) Six-year-old Youtube star brings his own toyline to Walmart. https://www.reuters.com/article/us-usa-toys/six-year-old-youtube-star-brings-his-own-toy-line-to-walmart-idUSKBN1KK1IQ. Accessed 9 Sept 2018

Richardson J, Schlegel K, Sallam R, Kronz A, Sun J (2021) Magic quadrant for analytics and business intelligence platforms, Gartner Group. Retrieved from: https://www.gartner.com/doc/reprints?id=1-1YOXON7Q&ct=200330&st=sb

Samuels M (2017) Big data case study: how UPS is using analytics to improve performance. ZDNet.com   https://www.zdnet.com/article/big-data-case-study-how-ups-is-using-analytics-to-improve-performance/. Accessed 9 Sept 2018

Saporito P (2015) Applied insurance analytics. Pearson Education, Upper Saddle River

Syvaniemi A (2015) Predictive analytics change internal audit practices. Blog newsletter. Retrieved at http://blog.houston-analytics.com/blog/predictive-analytics-change-internal-audit-practices

Troester M (n.d.) Big Data meets data analytics. SAS White Paper https://slidelegend.com/big-data-meets-big-data-analytics-sas_5a09b8481723ddee444f4116.html. Accessed 9 Sept 2018

van Rijmenjam M (2018) Why UPS spends $1 billion on data annually. https://datafloq.com/read/ups-spends-1.billion-big-dataannually/273. Accessed 9 Sept 2018

Walmart Staff (2017) 5 ways Walmart uses Big Data to help customers. https://blog.walmart.com/innovation/20170807/5-ways-walmart-uses-big-data-to-help-customers. Accessed 9 Sept 2018

Chapter 2
Know Your Data: Data Preparation

Chapter Learning Objectives
1. Differentiate between different data types.
2. Identify and explain the four basic SAS Enterprise Miner™ data roles.
3. Identify the methods used for handling missing values, outliers, and redundant data.
4. Using SAS Enterprise Miner™, prepare data for predictive modeling.

Once the business problem is identified, the hypotheses developed, and the data collected, the next step in the process is to analyze the data and prepare the data for predictive modeling. Most raw data are considered "dirty" or "noisy" because the data may have incomplete information, redundant information, outliers, or errors. Therefore, the data should be analyzed and "cleaned" prior to model development. Chapter 2 outlines the entire data analysis and preparation process. This is step 4 in the predictive analytics model. The chapter will begin with a description of the different categories of data followed by a review of the methods used for preparing the data. After a description of the data classifications and data preparation methods, the process will be reviewed step by step in SAS Enterprise Miner™ using the automobile insurance claim data set described in Appendix A.

2.1 Classification of Data

Once the data is collected, the data scientist or analyst needs to first understand the data. For example, how many records were collected, and how many variables are contained in the data? A variable or data item is an identifiable piece of data that can be measured or counted. The variable's values can appear as a number or text, which can be converted into number. Variables can have different properties or characteristics.

2.1.1 Qualitative Versus Quantitative

Before analyzing the data, it is important to distinguish between the two types of variables: qualitative and quantitative (or numeric). *Qualitative* or categorical variables are descriptive in nature, for example, the color of a sweater, the name of a city, or one's occupation. In SAS Enterprise Miner™, categorical variables are referred to as *class variables* (Fig. 2.1). *Quantitative* variables are numeric. They represent a measurable quantity. Quantitative data can also be further categorized as discrete or continuous. Variables are said to be *discrete* if the measurements are countable in a finite amount of time. They can only take on a limited number of values. For example, shoe size, the number of heads in a coin toss, and SAT scores are discrete variables.

Continuous variables can take on any value, usually within some range. When trying to determine if a variable is continuous versus discrete, check to see if the value can be divided and reduced to more exact measurements. For example, height, weight, and age are continuous variables. Age is considered continuous, and one can always break it down into smaller and smaller increments: 5 years, 2 months, 3 days, 6 h, 4 s, 4 milliseconds, 8 nanoseconds, etc. A person's age in years would be discrete.

2.1.2 Scales of Measurement

Scales of measurement represent the methods in which types of data are defined and categorized. The four scales of measurement are nominal, ordinal, interval, and ratio. The nominal and ordinal measurement scales are used to further describe class

Fig. 2.1 Variable classifications

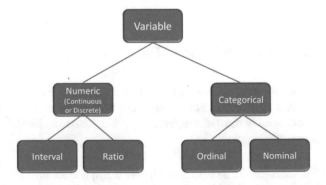

(qualitative/categorical) variables. Interval and ratio scales are generally used when referring to quantitative variables.

Nominal scales are used to identify or name qualitative values without any order. Examples of nominal scales include gender (male/female), hair color (blonde/brunette), or a student's major (marketing/finance/accounting). Nominal scales may be represented by numbers (1, male; 0, female). If so, these numbers are used to identify the object, and mathematical operations cannot be performed on the values. These values also do not represent a quantity or a measurement. For example, a zip code is a variable that would be a nominal scale measurement. Numbers represent its values. No meaningful data can be derived from adding, multiplying, or averaging zip codes.

Ordinal scales are where there is some natural relationship or rank order, such as a student's letter grade (A, B, C, D, F) or a shirt size (small, medium, large). Survey response that may range from strongly disagree to strongly agree is another example of ordinal data. For values measured on an ordinal scale, the difference between the values may not be equal. For example, the difference between a small beverage and a medium beverage may not be the same as the difference between a medium and a large beverage. So, it is the order of the values that is important, but the difference between each one is not known.

Interval scales characterize quantity, and the difference between the levels are equal. Examples of values that are measured with interval scales include the temperature in Fahrenheit or Celsius, a calendar year, an IQ score, or a student's SAT score. Let's look at temperature in Fahrenheit. The difference between 60 and 70 °F is the same as the difference between 50 and 60 °F. Interval values do not have a "true zero." For example, there is no such thing as having no temperature. A person cannot have a zero IQ or SAT score. Values measured on interval scales are ordered, and their differences may be meaningful; however evaluating a ratio or doubling its value is not meaningful. A 100 °F is not twice as warm as 50 °F.

Ratio scales have similar characteristics to the interval scales but also have a true (absolute) zero. The values have a rank order, and the intervals between values are equal. Doubling the values or calculating its ratios is meaningful. For example, doubling an object's weight is meaningful. It is twice as heavy. Height, weight, age, and temperature in Kelvin are all examples of ratio scale variables.

2.2   Data Preparation Methods

Data in its raw, original form is typically not ready to be analyzed and modeled. Data sets are often merged and contain inconsistent formats, missing data, miscoded data, incorrect data, and duplicate data. The data needs to be analyzed, "cleansed," transformed, and validated before model creation. This step can take a significant amount of time in the process but is vital to the process. Some common methods for handling these problems are discussed below.

2.2.1   Inconsistent Formats

Data in a single column must have consistent formats. When data sets are merged together, this can result in the same data with different formats. For example, dates can be problematic. A data column cannot have a date format as mm/dd/yyyy and mm/dd/yy. The data must be corrected to have consistent formats.

2.2.2   Missing Data

Missing data is a data value that is not stored for a variable in the observation of interest. There are many reasons that the value may be missing. The data may not have been available, or the value may have just been accidently omitted. When analyzing data, first determine the pattern of the missing data. There are three pattern types: missing completely at random (MCAR), missing at random (MAR), and missing not at random (MNAR). Missing completely at random occurs when there is no pattern in the missing data for any variable. Missing at random occurs when there is a pattern in the missing data but not on the primary dependent variables. An example of MAR would be the propensity of women not to tell their age. Missing not at random occurs when there is a pattern in the missing data that can affect the primary dependent variables. For example, if the study were on weight loss and heavier individuals were less likely to respond, this would affect the results.

In predictive modeling depending on the type of model being used, missing values may result in analysis problems. There are two strategies for dealing with missing values, listwise deletion or column deletion and imputation. Listwise deletion involves deleting the row (or record) from the data set. If there are just a few missing values, this may be an appropriate approach. A smaller data set can weaken the predictive power of the model. Column deletion removes any variable that contains missing values. Deleting a variable that contains just a few missing values is not recommended.

The second strategy and more advantageous is imputation. Imputation is changing missing data value to a value that represents a reasonable value. The common imputation methods are:

- Replace the missing values with another constant value. Typically, for a numeric variable, 0 is the constant value entered. However, this can be problematic; for example, replacing age with 0 does not make sense. For categorical variable, replace the missing value with a constant, for example, M, so it is known that the variable is missing. This works well when the missing value is completely at random (MCAR).
- Replace missing numeric values with the mean (average) or median (middle value in the variable). Replacing missing values with the mean of the variable is a common and easy method; plus it is likely to impair the model's predictability

as on average values should approach the mean. However, if there are many missing values for a particular variable, replacing with the mean can cause problems as more value in the mean will cause a spike in the variable's distribution and smaller standard deviation. If this is the case, replacing with the median may be a better approach.

- Replace categorical values with the mode (the most frequent value) as there is no mean or median. Numeric missing values could also be replaced by the mode.
- Replace the missing value by randomly selecting a value from missing value's own distribution. This is preferred over mean imputation however not as simple.

2.2.3 Outliers

An *outlier* is a data value that is an abnormal distance from the other data values in the data set. Outliers can be visually identified by constructing histograms, stem-and-leaf plots, or box plots and looking for values that are too high or too low. There are five common methods to manage the outliers:

1. Remove the outliers from the modeling data.
2. Separate the outliers and create separate models.
3. Transform the outliers so that they are no longer outliers.
4. Bin the data.
5. Leave the outliers in the data.

When the outliers are dropped, information in terms of the variability in data is lost. Before automatically removing outliers, examine the possible causes of the outliers and the affect the outlier may have on the analysis. If the outlier is due to incorrectly entered or measured data values, or if the outlier is outside the population of interest, it should be removed. If the outlier is likely to impair the model, in other words do more harm than good, the outlier should be removed. Some analysts will remove the outliers and create a separate model. Decision trees, discussed in Chap. 5, can aid in determining if the outliers are good predictors of the target variable.

Another method is to transform the data so that it is no longer an outlier. Some methods of handling the outlier include truncating, normalizing, or treating the outlier as missing data. If the outliers are too extreme such that they remain outliers after transformation, then it may be appropriate to bin the data. This method converts the numeric variable to a categorical variable through a process referred to as binning. This will be discussed further in Chap. 3.

There may be cases where outliers represent the population of data that is to be studied and should remain as is in the data set. There is no universal method for handling outliers; they should be considered within the context of the business problem.

2.2.4 Other Data Cleansing Considerations

Miscoded and incorrect values should be addressed. Identification of these errors can be done in a couple of ways depending on whether the variable is categorical or numeric. If categorical, frequency counts can aid in the identification of variables that are unusual and occur infrequently. In this case, categories can be combined. For numeric variables, they may be outliers. Duplicate values should be removed.

2.3 Data Sets and Data Partitioning

In predictive analytics to assess how well your model behaves when applied to new data, the original data set is divided into multiple partitions: training, validation, and optionally test. Partitioning is normally performed randomly to protect against any bias, but stratified sampling can be performed. The training partition is used to train or build the model. For example, in regression analysis the training set is used to fit the linear regression model. In neural networks, the training set is used to obtain the model's network weights. After fitting the model on the training partition, the performance of the model is tested on the validation partition. The best-fit model is most often chosen based on its accuracy with the validation data set. After selecting the best-fit model, it is a good idea to check the performance of the model against the test partition which was not used in either training or in validation. This is often the case when dealing with big data. It is important to find the correct level of model complexity. A model that is not complex enough, referred to as *underfit*, may lack the flexibility to accurately represent the data. This can be caused by a training set that is too small. When the model is too complex or *overfit*, it can be influenced by random noise. This can be caused by a training set that is too large. Often analysts will partition the data set early in the data preparation process.

2.4 SAS Enterprise Miner™ Model Components

The auto insurance claim data set will be used to walk through the process of data preparation using SAS Enterprise Miner™. Recall that the goal is to create a predictive model that will identify potential fraudulent claims. In Chap. 1, the business problem and hypotheses were identified.

To create a predictive model in SAS Enterprise Miner™ requires the creation of four components:

1. *Project*—contains the diagrams, data sources, and the library for the data. Generally, a separate project is created for each problem that is trying to be solved.
2. *Diagram*—This is the worksheet where you build your model and determines the processing order and controls the sequence of events for preparing the predictive model.
3. *Library*—is a pointer to the location of the stored data files.
4. *Data Source*—contains both the data and the metadata that defines and conFig.s an input data set.

The SAS Enterprise Miner™ data preparation steps are:

1. Create the first three model components: project, diagram, and library.
2. Import the data and convert to a SAS file.
3. Create the data source.
4. Partition the data.
5. Explore the data.
6. Address missing values.
7. Address outliers.
8. Address categorical variables with too many levels.
9. Address skewed distributions (addressed in Chap. 3).
10. Transform variables (addressed in Chap. 3).

2.4.1   Step 1. Create Three of the Model Components

Creating the Project File
When using SAS Enterprise Miner™, the work is stored in projects. When SAS Enterprise Miner™ was first launched, the Welcome screen appears (Fig. 2.2).

Upon selecting New Project, the create New Project window opens (Fig. 2.3). In this window enter the name of the project, and indicate the directory location where the project will be stored. Enter "Claim Fraud" for the project name. Then click *Browse*, and indicate the location where the project files will be stored. Then click *Next*. The step 2 of 2 screen indicates the project name and the directory. Click *Finish*.

The software then creates the project files, and the SAS Enterprise Miner™ Project Work Area screen is displayed.

Fig. 2.2 Welcome screen

Fig. 2.3 Create New Project window

Create the Diagram
To create a diagram from the File menu, select New—Diagram or right click Diagrams in the Project Panel and select Create Diagram. Name the Diagram Create SAS Datasets (Fig. 2.4).

Create the Library
In order to access the automobile insurance claim data set, a SAS library must be created. The library creates a shortcut name to the directory location of the stored SAS data files, and SAS creates a pointer to that directory. To create a SAS library, click on the File menu—New—Library. The library wizard then opens. Be sure that Create a New Library is selected and click Next. In Step 2 Window (Fig. 2.5), enter

Fig. 2.4 Create data set diagram

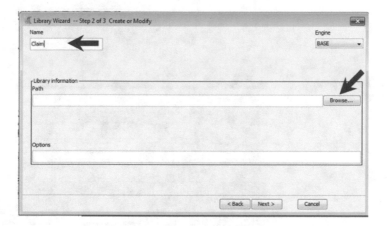

Fig. 2.5 Create a library wizard—Step 2

Claim as the library name, and then browse to the directory where the SAS data file will be stored. Click Next and Finish.

2.4.2   Step 2. Import an Excel File and Save as a SAS File

This next step takes an Excel workbook (.xlsx) and converts the file to a SAS Enterprise Miner™ data set (.bdat). The SAS data file contains two components: the first is the data itself which is the records or observations. It is analogous to an Excel spreadsheet containing rows and columns. The rows contain the records, while the columns contain the variables. The second component is metadata. Metadata defines to SAS the roles and levels of each variable.

To import the data, drag the Import node (Fig. 2.6) from the Sample tab onto the Create SAS Datasets Process Flow (Fig. 2.7). A node is how you implement a

Fig. 2.6   Import node

Fig. 2.7   Process flow
diagram with Import node

Fig. 2.8   Import node
properties

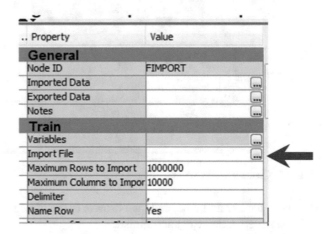

specific process in SAS Enterprise Miner™. Each step in the predictive model is
implemented with a node that defines that process (e.g., there is one for data partition, impute, etc.)

With the Import node selected, click on the ellipsis by the Import File property
in the Train group (Fig. 2.8). The File Import window (Fig. 2.9) opens; click Browse
and then locate the claim fraud.xls Microsoft Excel data file on the computer. To run
the specific node, right-click on the node and select Run, or to run the entire model,
click on the Run icon (Fig. 2.10). A green check mark in the corner of the node
indicates that SAS Enterprise Miner™ successfully ran the node (Fig. 2.11).

Fig. 2.9 File Import window

Fig. 2.10 Run icon

Fig. 2.11 File Import

Fig. 2.12 Save Data node

To save the imported Excel file as a SAS file, drag the Save Data node (Fig. 2.12) from the Utility tab to the Process Flow Diagram. To connect the Save Data node, move the mouse to the right side of the File Import node until the cursor becomes a pencil, and click and drag the pencil to the left-hand side of the File Import node (Fig. 2.13).

On the Save Data node properties, set the Filename Prefix property to Claim_ Data. To reference a SAS data set, it must be stored in the SAS library defined to the project. Click the ellipsis by the SAS Library Name property, and select the newly created SAS Library Claim (Fig. 2.14). Run the node and then click OK.

The rest of the model can be continued on the same process flow diagram. However, to better organize the model, a second diagram will be created. Create a second diagram named Data Preparation and Analysis (Fig. 2.15).

Fig. 2.13 Process flow diagram with Save Data node

Fig. 2.14 Save Data node
properties

Fig. 2.15 Data Preparation
and Analysis diagram

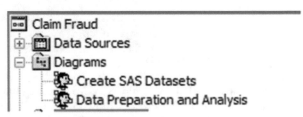

2.4.3   Step 3. Create the Data Source

The next step is to create a new data source. To use the auto insurance claim data set, a data source must be created. To create the data source, select the File menu—New—Data Source, or right-click on Data Sources in the Project Panel. The Data Source Wizard opens. A SAS Table is automatically selected as the source. Click Next and then click the Browse button, and navigate to the claim library and to the automobile insurance claim data file and click OK (Fig. 2.16) and then Next.

Metadata as used in SAS Enterprise Miner™ defines the roles and levels of each variable.

Fig. 2.16 Data Source Wizard—SAS Table

Step 3 of 3 of the Data Source Wizard displays information on the data set including the number of variables and the dates the data set was created and last modified. Click Next. At this point metadata needs to be defined. The metadata contains information on the data set including its name and library path. It also stores the variable roles and data types. If the Basic Option is selected, SAS will determine the initial roles and level based upon the variable type and format. If the variable type is numeric, its measurement scale will be interval, and all class variables or categorical variables will have a measurement scale of nominal. If the Advanced Option is selected, SAS sets the measurement scale of any numeric variable with less than 21 unique values to nominal. In the Advanced Option, the variables' measurement scales can be customized.

Data Roles
The data variables in SAS Enterprise Miner™, each has to be assigned a role. There are four frequently used data roles: id, input, target, and rejected. The "id" role is the primary key and is an identifier used to identify a unique row of data. Generally, the input variables are the independent variables, and the targets are the dependent variables hereafter referred to as the target variable. Keep in mind, the input variable is the variable whose change isn't affected by any other variable in the model, whereas the target variable is what is being measured or predicted in the model. Rejected variables are variables in the data set that will not be included for analysis. Variables can be rejected for several reasons including if they are duplicates and have no influence on the target variable or if you have too many variables. The automobile insurance claim data set's data dictionary includes the data type, roles, and definitions that are given in Appendix A. Notice that the variable state is rejected because it defines the same data term as state_code.

Data Levels
SAS Enterprise Miner™ calls scales of measurement levels. It does not use ratio as a level. In SAS Enterprise Mine™, there are five levels:

1. The binary levels have only two possible values. For example, on/off or yes/no.
2. The interval level as defined above in Sect. 2.1.
3. The nominal level as defined above in Sect. 2.1.

Fig. 2.17 Data Source Wizard—Step 5 Setting the Metadata Roles

4. The ordinal level as defined above in Sect. 2.1.
5. Unary a variable that contains a single, discrete value and is rarely used.

To change the roles and levels, select the Advanced Metadata Advisor and then click Next. Use the column metadata table editor to conFig. the variable roles as shown in Fig. 2.17. To change an attribute's role, click on the value of that attribute, and select a new value from the drop-down list that appears. For this project, set the target variable to Fraudulent_Claim. This is the dependent variable; it is a binary (yes/no) variable that specifies if the claim is fraudulent. Click on the Input Role for Fraudulent_Claim, and change it to Target. The Claimant_Number is an ID variable. Change its role to ID. The state variable will not be used in the model and is therefore rejected. It is common in predictive analytics to have many variables in a Data Source, but that does not mean they all need to be considered.

After the variable role assignments have been set for the data set, the Decision Processing window opens (Fig. 2.18). At this step, the target specifications can be assigned. Profits or costs for each possible decision, prior probabilities, and cost functions can be set. The auto insurance claim example does not assign any monetary costs for correct or wrong decisions; the goal is to identify fraudulent claims. So, make sure No is selected and click Next.

Figure 2.19 shows steps 7 of 9 of the Data Source. A sample data set is not needed so make sure No is selected and click Next

For Step 8 click Next (Fig. 2.20) and then click Finish.

Fig. 2.18 Data Source Wizard—Step 6 Decision Processing

Fig. 2.19 Data Source Wizard—Step 7 Sample Data Set

Fig. 2.20 Data Source Wizard Step

Fig. 2.21 Project Panel

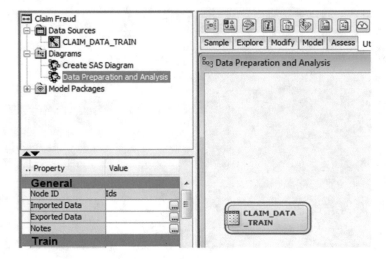

Fig. 2.22 Data Source node

The project panel will now display the new data source (Fig. 2.21). The extension _Train is added automatically to the end of the data source name.

The last step is to drag the Data Source from the Project Panel to the Diagram located in the workspace. This will create a Data Source node in the process flow (Fig. 2.22). Run the CLAIM_DATA_TRAIN node and view the results. The results show there are 5001 observations in the data set and a list of the variables are provided in the Variables window (Fig. 2.23).

2.4.4   Step 4. Partition the Data Source

Partitioning the data set divides the data set into two or three parts: train, validation, and testing (optional). The training partition is used to build the model; the validation partition is created to validate or check the accuracy of the model; and the optional test partition tests the model.

Fig. 2.23 Data Source node results

Fig. 2.24 Data
Partition node

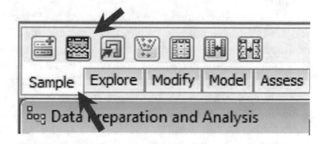

To create the partition, click on the Sample tab, and then drag the Data Partition node onto the process flow (Fig. 2.24). Connect the Data Partition node to the CLAIM_DATA_TRAIN node (Fig. 2.25).

Select the Data Partition node, and set the partition allocations as shown in Table 2.1.

Sixty percent of the data set will be used to train or create the model, and 40% will be used to validate it. We will not test the data. Note: the sum of the allocations must be 100 (i.e., 100%). There is no perfect cross validation or partition size. It depends on the problem and the data set size. The smaller the data set, the larger the training set needs to be. For example, if you have a very small data set, you might need to have an 80/20 split. It is common to begin with a partition size of 60/40. For large data sets, a common partition size is 50/30/20. Run the model.

Fig. 2.25 Process flow diagram with Data Partition

Table 2.1 Data Partition allocations

Data Set Allocations	
Training	60.0
Validation	40.0
Test	0.0

2.4.5 Step 5. Data Exploration

In SAS Data Enterprise Miner™, the data preparation and data cleansing process begin by looking at the data. The StatExplore and the Graph Explore nodes provide useful information about the data. The StatExplore node provides descriptive statistics and is used to aid in identifying missing values. The Graph Explore node provides a variety of charts of the data including histograms, stem-and-leaf plots, and box plots.

The StatExplore node in SAS can be used to explore the data and locate those variables that have missing values. Once the missing values have been identified, the Impute node can be used to handle the missing value. To examine the auto insurance claim data, drag the StatExplore node (Fig. 2.26) to the process flow, and connect it to the Claim_Train_DataSet (Fig. 2.27). Run the model and view the results.

Scrolling down through the Output Window notices that the Education attribute is missing 11 values (Fig. 2.28).

To further explore the data, click on the Explore tab, and drag the Graph Explore node (Fig. 2.29) onto the process flow diagram (Fig. 2.30). Connect the Graph Explore node to the CLAIM_DATA_TRAIN node, run the model, and then view the results.

Right-click on the Graph Explore node and select Edit Variables. Click on the Role column title to sort by Role. To select all of the inputs, click on the Marital_Status variable name, hold the shift down, and click on Claim_Amount (Fig. 2.31). Click Explore.

A histogram of all of the inputs is now displayed (Fig. 2.32). Looking at the histograms, it appears that the Income variable has an outlier.

2.4.6 Step 6. Missing Data

To overcome the obstacle of missing data, missing values can be imputed before the model is fitted. Recall that imputation replaces the missing values with substitute values. SAS Enterprise Miner™ can perform both listwise deletion and imputation

Fig. 2.26 StatExplore node

Fig. 2.27 Process flow diagram with StatExplore node

```
37   Class Variable Summary Statistics
38   (maximum 500 observations printed)
39
40   Data Role=TRAIN
41
42                                        Number
43   Data                                 of                        Mode                     Mode2
44   Role    Variable Name      Role      Levels  Missing  Mode      Percentage  Mode2       Percentage
45
46   TRAIN   Education          INPUT     6       11       College   29.83       Bachelor    29.81
47   TRAIN   Employment_Status  INPUT     5       0        Employed  62.51       Unemployed  25.39
48   TRAIN   Gender             INPUT     2       0        M         50.61       F           49.39
49   TRAIN   Marital_Status     INPUT     3       0        Married   58.59       Single      26.41
50   TRAIN   State_Code         INPUT     5       0        IA        30.91       MO          28.91
51   TRAIN   Vehicle_Class      INPUT     6       0        Four-Door Car  50.69  Two-Door Car  20.58
52   TRAIN   Vehicle_Size       INPUT     3       0        Midsize   70.45       Compact     19.30
53   TRAIN   Fraudulent_Claim   TARGET    2       0        N         93.86       Y           6.14
```

Fig. 2.28 StatExplore Output Window

Fig. 2.29 Graph Explore node

Fig. 2.30 Process flow diagram with Graph Explore node

Fig. 2.31 Graph Explore—variable selection

Fig. 2.32 Histograms of input variables

automatically. The Impute node is used to deal with missing values. Table 2.2 lists the imputation methods available in SAS Enterprise Miner™.

Table 2.2   SAS Enterprise Miner™ imputation methods (Brown 2015)

Class (categorical) variables	Interval (numeric) variables
Count	Central tendency measures
Default constant value	(mean, median, and midrange)
Distribution	Distribution
Tree (only for inputs)	Tree (only for inputs)
Tree surrogate (only for inputs)	Tree surrogate (only for inputs)
	Mid-minimum spacing
	Tukey's Biweight
	Huber
	Andrew's wave
	Default constant

By default, SAS Enterprise Miner™ uses a sample from the training data set to select the values for replacement. Mean is the default for interval variables. The mean is calculated from the training sample. Count is the default method for missing values of binary, nominal (class), or ordinal variables (class). With count imputation, the missing values are replaced with the most commonly occurring level of the corresponding variable in the sample. -

The mean imputation method is not always recommended because it can artificially reduce the variability of the data. It also changes the magnitude of correlations between imputed variables and other variables. In addition to mean, median is a common imputation method for interval variables. Median imputation replaces the missing values with the median of the non-missing values. The median is less sensitive to extreme values than the mean. The median is useful for replacing values in skewed distributions.

To handle the missing values, click on the Modify tab (Fig. 2.33), and then drag the Impute node to the process flow diagram (Fig. 2.34).

Change the default input method for the interval variables in the train data set to median (Fig. 2.35). Then run the model and view the results (Fig. 2.36).

In SAS Enterprise Miner™, the Impute node creates a new variable with the new replacement values for the missing data. The Impute node defaults so that the original data values in the original data set are not overwritten. Instead a new variable is created that contains the imputed values. Imputed variables can be identified by the prefix IMP_ added to the original variable name. Notice in the Output Window that the IMP_Education variable was added to the data set (Fig. 2.32).

2.4.7   Step 7. Handling Outliers

In SAS Enterprise Miner™, the *Filter* node can be used to both identify and exclude outliers of class or interval variables. It also can be used to perform general filtering of the data set. Filtering extreme values from the training data tends to produce

Fig. 2.33 Impute node

Fig. 2.34 Process flow
diagram with Impute node

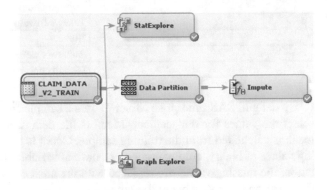

Fig. 2.35 Impute node
settings

```
36    Imputation Summary
37    Number Of Observations
38
39                                                                              Number of
40    Variable    Impute      Imputed      Impute               Measurement        Missing
41     Name       Method      Variable      Value    Role          Level    Label  for TRAIN
42
43    Education   COUNT    IMP_Education   College   INPUT       NOMINAL  Education      5
44
```

Fig. 2.36 Impute Results

better models. However, recall that there may be business problems where it is important to consider the impact of outliers. By default, the Filter node ignores target and rejected variables. The available filtering methods for class variables include:

- Rare values (count): This filter method drops variables that have a count less than the level that is specified in the Minimum Frequency Cutoff property.
- Rare values (percentage): This is the default setting; it drops rare variables that occur in proportions lower than the percentage specified in the Minimum Cutoff for Percentage property. The default percentage is set to 0.01%.
- None: No class variable filtering is performed.

The Normalized Values property default setting is Yes. Normalizing class variables left-justifies them, ignores capitalization differences, and truncates them to 32. If the property is set to No, the normalization of class variable values before filtering does not occur.

For interval variables, the Interval Variables property can be set to different filter methods for individual variables instead of applying the default filtering method to all of the interval variables. The available filtering methods for interval input variables include:

- Mean absolute deviation (MAD): Removes values that are more than n deviations from the median. The threshold value, n, is specified in the Cutoff for MAD property.
- User-specified limits: Sets a lower and upper limit. The limits are specified in the Variable table in the Filter Lower Limit and Filter Upper Limit columns.
- Extreme percentiles: Filters the values that are in the top and bottom percentile. The upper and lower bound values are specified in the Cutoff Percentiles for Extreme Percentiles property.
- Modal center: Removes values that are more than n spacings from the modal center. The threshold value is specified in the Cutoff Percentiles for Modal Center property.
- Standard deviations from the mean: This is the default. It filters values that are greater than or equal to n standard deviations from the mean. The threshold is set in the Cutoff Percentiles for Standard Deviations property.
- None: There is no filter applied.

The Keep Missing Values properties for interval and class variables determine if variables that are missing data are kept. Set the Keep Missing Values property of the Filter node to No to filter out observations that contain missing values. The default setting for the Keep Missing Values property is Yes.

Recall that the results of the Graph Explore node indicated that the income variable had outliers. To filter out the outliers, add a Filter node (Fig. 2.37). From the Sample tab, drag a Filter node to the process flow diagram, and connect it to the Impute node as shown in Fig. 2.38.

Under the Train grouping, set the Tables to Filter to All Data Sets. Since Income is the only variable to be filtered, the filter settings will be set individually by using

Fig. 2.37 Filter node

Fig. 2.38 Process flow diagram with Filter node

Fig. 2.39 Filter node with
User-Specified Limits

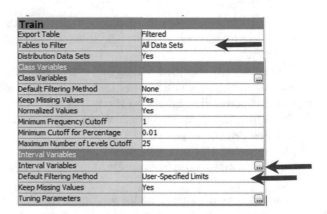

the Interval Variables property. In the Filter properties window, click the ellipsis by
the Interval Variables property, and set the minimum income to zero and the maximum income to 103,677.78 (i.e., 3* 34,559.26) (Figs. 2.39 and 2.40). The 103,677.78
value is approximately three standard deviations from the mean. For a normal distribution, approximately 99.7% of the values are within three standard deviations.
Therefore, using three standard deviations is reasonable. Be sure the default filtering
method is set to User-Specified Limits.

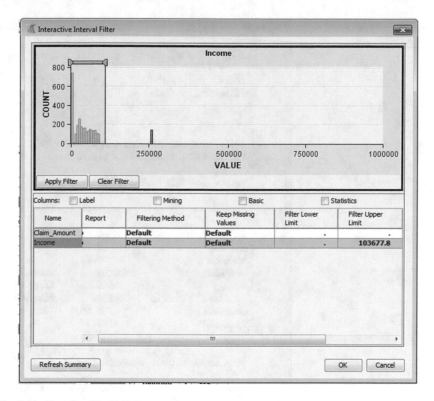

Fig. 2.40 User-Specified Limits

Click OK, run the filter again, and view the results (Fig. 2.41).

From the Output Window results, one observation from the train partition and two observations from the validate data partition were filtered. Scrolling further down in the results, the maximum income value in the train partition is $99,981 as opposed to the original maximum of $933,288.

2.4.8 Step 8. Categorical Variables with Too Many Levels

Categorical or class variables with too many different values can reduce the performance of the model. A common example of this is the zip code variable. One method of dealing with this is to combine values into similar groups. For example, zip code can be combined at the city or state level. Levels can also be combined based upon their frequency. In SAS Enterprise Miner™, the replacement editor (within the properties) of the replacement node can be used to combine groups and establish different levels for the groups. The Filter node can be used to set default minimum frequencies and minimum number of levels (Fig. 2.42). The Interactive Class Filter (Click on the Ellipsis by Class Variables) can be used to manually establish these

Fig. 2.41 User-specified
filter results

```
 Output
   44    Data
   45    Role          Filtered      Excluded      DATA
   46
   47    TRAIN            2997            1          2998
   48    VALIDATE         2001            2          2003
   49
   50     |
   51
   52    Statistics for Original and FILTERED Data
   53    (maximum 500 observations printed)
   54
   55    Data Role=TRAIN Variable=Income
   56
   57    Statistics                  Original      Filtered
   58
   59    Non Missing                  2998.00       2997.00
   60    Missing                         0.00          0.00
   61    Minimum                         0.00          0.00
   62    Maximum                    933288.00      99981.00
   63    Mean                        38326.89      38028.27
   64    Standard Deviation          34559.26      30451.78
   65    Skewness                        5.97          0.28
   66    Kurtosis                      148.50         -1.11
   67
   68
   69    Data Role=VALIDATE Variable=Income
   70
   71    Statistics                  Original      Filtered
   72
   73    Non Missing                  2003.00       2001.00
   74    Missing                         0.00          0.00
   75    Minimum                         0.00          0.00
   76    Maximum                  15967801.00      99960.00
   77    Mean                        45776.10      37717.00
```

Fig. 2.42 Filter node—
class variables

Class Variables	
Class Variables	...
Default Filtering Method	None
Keep Missing Values	Yes
Normalized Values	Yes
Minimum Frequency Cutoff	1
Minimum Cutoff for Percentage	0.01
Maximum Number of Levels Cutoff	25

limits (Fig. 2.43). Click Generate Summary to view the histograms in the Interactive Class Filter.

The histograms of the class variables in the Interactive Class Variable Filter and the output from the StatExplore Node (Fig. 2.44) indicate that the number of levels for the class variables is reasonable.

```
40   Data Role=TRAIN
41
42                                       Number
43   Data                                  of                                        Mode                          Mode2
44   Role      Variable Name      Role    Levels   Missing   Mode        Percentage   Mode2        Percentage
45
46   TRAIN     Education          INPUT     6        11      College        29.83     Bachelor        29.81
47   TRAIN     Employment_Status  INPUT     5         0      Employed       62.51     Unemployed      25.39
48   TRAIN     Gender             INPUT     2         0      M              50.61     F               49.39
49   TRAIN     Marital_Status     INPUT     3         0      Married        58.59     Single          26.41
50   TRAIN     State_Code         INPUT     5         0      IA             30.91     MO              28.91
51   TRAIN     Vehicle_Class      INPUT     6         0      Four-Door Car  50.69     Two-Door Car    20.58
52   TRAIN     Vehicle_Size       INPUT     3         0      Midsize        70.45     Compact         19.30
53   TRAIN     Fraudulent_Claim   TARGET    2         0      N              93.86     Y                6.14
54
```

Fig. 2.44 StatExplore output

Fig. 2.43 Interactive class filter

2.5   Chapter Summary

An understanding of variable data types and roles is necessary before any analysis
can be started. Class variables are descriptive and are categorical, whereas quantitative variables are numeric. Class variables have a measurement scale of either

nominal (unordered) or ordinal (ordered). Numeric variables have a measurement scale of either interval or ratio. Interval variables do not have a "true zero" point, whereas ratio scale variables have a true zero point.

When setting up the metadata for modeling, you need to set the data types or levels and the roles of the variables or attributes. There are four commonly used roles: id, target (dependent variable), input (independent variable), and rejected. The rejected variables are not included in the model.

In SAS Enterprise Miner™, the first step in building a predictive analytics model requires the creation of four components: a project, diagram, library, and a data source. The project contains the diagrams, data sources, and the library for the data. The library is a pointer to the location of the stored SAS data files. The data source contains both the data and the metadata. To be able to measure how well the model behaves when applied to new data, the original data set is divided or partitioned into multiple partitions: training, validation, and test.

Part of the data preparation process is to handle missing values and outliers. The Imputation node is used to process missing values. Common methods for replacing the missing values are with the mean, median, most common value, or some constant value, or simply remove the record with the missing value. The Filter node is used to both identify and exclude outliers of class or interval variables. The filter can be set manually or based upon other factors such as the number of standard deviations from the mean.

If the categorical variables have too many levels, this can affect the performance of the model. Consider reducing the number of levels by combining them. The Filter node can be used to assist in this process.

Discussion Questions

1. Describe the different variable types and the measurement scales. Give examples of each.
2. Discuss the different methods for handling missing values and outliers.
3. What are some of the data problems you should adjust for before modeling your data?
4. Why is data partitioned?
5. What are some methods for filtering data? Provide examples of each.

Reference

Brown I (2015) Data exploration and visualisation in SAS Enterprise Miner, SAS Forum United Kingdom. Available via https://www.sas.com/content/dam/SAS/en_gb/doc/other1/events/sasforum/slides/day2/I.%20Brown%20Data%20Exploration%20and%20Visualisation%20in%20SAS%20EM_IB.pdf. Accessed on 9 Sept 2018

Chapter 3
What Do Descriptive Statistics Tell Us

Chapter Learning Objectives

1. Perform hypothesis testing and identify the two types of errors that can occur.
2. Describe the importance of central tendency, variation, and shape for numerical variables and how it can affect your predictive model.
3. Compare and contrast data distributions and how to correct for skewness and kurtosis.
4. Calculate the covariance and the coefficient of correlation.
5. Evaluate ANOVA results and describe when it is appropriate to use.
6. Evaluate Chi-square results and describe when it is appropriate to use.
7. Identify the various methods for evaluating the fit of a predictive model.
8. Identify and describe the methods for variable reduction.
9. Describe stochastic models.

Prior to analyzing any data set, it is important to first understand the data. Descriptive statistics presents data in a meaningful way for the purpose of understanding what if anything will need to be done to the data to prepare it for analysis. There are many statistical tests that can be utilized. This chapter focuses on a review of descriptive statistical analysis that is used to prepare and support predictive analytics. It reviews methods to ensure that the data is prepared for analysis as well as methods for combing or reducing variables to improve the results of predictive analytics.

3.1   Descriptive Analytics

Descriptive analytics is the first stage of data analysis. It provides a summary of historical data that may indicate the need for additional data pre-processing to better prepare the data for predictive modeling. For example, if a variable is highly skewed, the variable may need to be normalized to produce a more accurate model.

This chapter covers descriptive analysis methods and reviews several other basic statistical methods to further process the data. Descriptive analytics summarizes a data set, which can be either a representation of the entire population or just a sample. Organizations performing predictive analytics may use their own data; therefore, they have access to the entire population, whereas marketing analytics is often working with sample data (survey responses, tweets, purchased data, etc.)

Descriptive statistics are broken down into measures of central tendency and measures of variability and shape. Measures of central tendency include the mean, median, and mode. Measures of variability include the standard deviation, variance, range, and the kurtosis and skewness.

This chapter reviews descriptive concepts and several statistical methods that will be utilized in developing and analyzing a predictive model. Correlation, hypothesis testing, the Chi-square test, principal component analysis (PCA), ANOVA, model-fit statistics, and stochastic models are briefly summarized. The chapter also covers several techniques for variable reduction, including principal component analysis and variable clustering.

3.2   The Role of the Mean, Median, and Mode

The *central tendency* is the extent to which the values of a numeric variable group cluster around a typical or central value. The three measures we will look at are the arithmetic mean, the median, and the mode. The *mean* is the most common measure of central tendency. It is the sum of all the values divided by the number of values. Although it is the preferred measure, extreme values or outliers affect it.

The median is another measure of central tendency. It is the middle value in an ordered set of observations. It is less sensitive to outliers or extreme values. If the number of observations is odd, the median is simply the middle number. If the number of observations is even, the median is the average of those two values on either side of the center. The median is not influenced by outliers. The third measure of central tendency is the mode. The mode is the value that occurs most often. Figure 3.1 shows an example of the calculation of the mean, median, and mode. Which measure should you choose? Unless extreme values or outliers exist, the mean is generally the preferred measure. The median is used when the data is skewed; there are a small number of observations or working with ordinal data. The mode is rarely used. The only situation in which the mode would be preferred is when describing categorical or class variables. Often the greatest frequency of

Observations: 3 6 17 5 3   8 5 3 22 4   n=10

Ordered Observations: 3 3 3 4 5 5 6 8 17 22

Mean: (3+ 6 + 17 + 5 + 3 + 8 + 5 + 3 + 22 + 4)/10 = $\boxed{7.6}$

Median: 3 3 3 4 5 5 6 8 17 22

Average of two middle values: (5 + 5)/2 = $\boxed{5}$

Mode: $\boxed{3}$ (The value of 3 occurs 3 times in the dataset)

Fig. 3.1 Calculating measures of central tendency example

$$S^2 = \frac{\Sigma(x - \bar{x})^2}{n-1}$$

Where x = the individual observation
Where $\bar{x}$ = the mean of the observation
n = the number of observations

Fig. 3.2 Variance formula

observations is the measure of interest. Frequently both the mean and the median are reported. For example, the median and mean house prices may be reported for a region.

3.3   Variance and Distribution

Measures of variation provide information on the spread or variability of the data set. Some of the measures of variation include the range, the sample variance, the sample standard deviation, and the coefficient of variation. The *range* is the easiest measure to calculate. It is the difference between the highest and the lowest value. The range does not provide any information about the distribution of the data. The range is also sensitive to outliers. A common measure of variation is the sample variance. The *sample variance* is the average of the squared deviations of each observation from the mean. Figure 3.2 shows the formulas for the sample standard deviation, usually denoted as S^2.

The *standard deviation* is the square root of the variance and is in the same units of measurement as the original data. Figure 3.3 shows an example calculation. The more the data is spread out, the greater the range, variance, and standard deviation. The more the data is concentrated, the smaller the range, variance, and standard deviation.

Returning to the automobile insurance claim fraud example, the Output Window of the StatExplore node, created in Chap. 2, provides many of the descriptive statistics. Figure 3.4 shows the summary statistics on the class variables. The most common value or mode for the categorical variable Education is College, which makes

Observations: 3 6 17 5 3 8 5 3 22 4   n=10

Range: 22-3 = (19)

Mean = 7.6

Variance: $((3-7.6)^2 + (6-7.6)^2 + (17-7.6)^2 + (5-7.6)^2 + (3-7.6)^2 + (8-7.6)^2 + (5-7.6)^2 + (3-7.6)^2 + (22-7.6)^2 + (4-7.6)^2)/(10-1) = (43.16)$

Standard Deviation: $\sqrt{43.16} = (6.57)$

Fig. 3.3 Calculating the variance and distribution

```
37
38    Class Variable Summary Statistics
39    (maximum 500 observations printed)
40
41    Data Role=TRAIN
42
43                                           Number
44    Data                                     of                        Mode                         Mode2
45    Role      Variable Name      Role      Levels    Missing   Mode     Percentage   Mode2          Percentage
46
47    TRAIN     Education          INPUT       6         11      College     29.83     Bachelor         29.81
48    TRAIN     Employment_Status  INPUT       5          0      Employed    62.51     Unemployed       25.39
49    TRAIN     Gender             INPUT       2          0      M           50.61     F                49.39
50    TRAIN     Marital_Status     INPUT       3          0      Married     58.59     Single           26.41
51    TRAIN     State_Code         INPUT       5          0      IA          30.91     MO               28.91
52    TRAIN     Vehicle_Class      INPUT       6          0      Four-Door Car 50.69   Two-Door Car     20.58
53    TRAIN     Vehicle_Size       INPUT       3          0      Midsize     70.45     Compact          19.30
54    TRAIN     Fraudulent_Claim   TARGET      2          0      N           93.86     Y                 6.14
55
```

Fig. 3.4 StatExplore Output Window—class variable summary statistics

```
71    Interval Variable Summary Statistics
72    (maximum 500 observations printed)
73
74    Data Role=TRAIN
75
76                                  Standard      Non
77    Variable      Role     Mean   Deviation   Missing   Missing   Minimum    Median     Maximum    Skewness   Kurtosis
78
79    Claim_Amount  INPUT  787.7633  655.9633    5001        0      189.8684   577.3521   7422.852   2.922369   12.62942
80    Income        INPUT  41310.45  227690.4    5001        0         0        34621    15967801   68.48667   4790.542
81
82
```

Fig. 3.5 StatExplore Output Window—interval variable summary statistics

up 29.83% of the observations. The second most frequently occurring observation for the education variable was Bachelor with 29.81% of the observations.

Figure 3.5 shows the summary statistics on the interval variables. The mean claim amount was $787.7633 with a standard deviation of $655.9633. The median (middle value) claim amount was $577.3521.

The StatExplore output also provides summary statistics by the target variable (Fraudulent_Claim) for both categorical (Fig. 3.6) and interval variables (Fig. 3.7). The mode for the input variable, Education, and with a target variable, Fraudulent_ Claim value of no (N), was College, whereas the mode for the input variable Education with a target variable of yes (Y) for Fraudulent_ Claim was Bachelor.

Figure 3.7 shows that the median and mean for the target variable Fradulent_ Claim that were not fraudulent (N) were $581.0631 and $792.7236, respectively,

```
84   Class Variable Summary Statistics by Class Target
85   (maximum 500 observations printed)
86
87   Data Role=TRAIN Variable Name=Education
88
89                           Number
90                Target     of                      Mode                        Mode2
91   Target       Level      Levels   Missing  Mode   Percentage  Mode2      Percentage
92
93   Fraudulent_Claim   N      6        11      College    29.91    Bachelor    29.83
94   Fraudulent_Claim   Y      5         0      Bachelor   29.64    College     28.66
95   _OVERALL_                 6        11      College    29.83    Bachelor    29.81
96
97
98   Data Role=TRAIN Variable Name=Employment_Status
99
100                          Number
101               Target     of                      Mode                        Mode2
102  Target       Level      Levels   Missing  Mode   Percentage  Mode2      Percentage
103
104  Fraudulent_Claim   N      5         0      Employed   62.89    Unemployed   24.95
105  Fraudulent_Claim   Y      5         0      Employed   56.68    Unemployed   32.25
106  _OVERALL_                 5         0      Employed   62.51    Unemployed   25.39
107
```

Fig. 3.6 Output Window—class variable summary statistics by class target

```
165  Interval Variable Summary Statistics by Class Target
166  (maximum 500 observations printed)
167
168  Data Role=TRAIN Variable=Claim_Amount
169
170              Target                      Non                        Standard
171  Target      Level   Median  Missing  Missing  Minimum   Maximum   Mean      Deviation  Skewness  Kurtosis  Role   Label
172
173  Fraudulent_Claim  N   581.0631   0     4694   189.8684  7422.852  792.7236  660.5073   2.948786  12.86009  INPUT  Claim_Amount
174  Fraudulent_Claim  Y   534.5681   0      307   200.4351  3345.179  711.9203  577.7544   2.208033  4.72192   INPUT  Claim_Amount
175  _OVERALL_             577.3521   0     5001   189.8684  7422.852  787.7633  655.9633   2.922369  12.62942  INPUT  Claim_Amount
176
177
178  Data Role=TRAIN Variable=Income
179
180              Target                      Non                        Standard
181  Target      Level   Median  Missing  Missing  Minimum   Maximum   Mean      Deviation  Skewness  Kurtosis  Role   Label
182
183  Fraudulent_Claim  N    34990     0     4694      0    15967801  41798.41  234882.6   66.4627   4506.54   INPUT  Income
184  Fraudulent_Claim  Y    29735     0      307      0      98701   32849.56  30482.01   0.417743  -1.02688  INPUT  Income
185  _OVERALL_            34621     0     5001      0    15967801  41310.45  227690.4   68.48667  4790.542  INPUT  Income
186
```

Fig. 3.7 Output Window—interval variable summary statistics by class target

whereas the median and mean for the target variable Fraudulent_Claim that were fraudulent (Y) waere $534.5681 and $711.9203, respectively.

The output of the StatExplore node also provides additional statistics on the interval and class (categorical) variables. They can be viewed by selecting View—Summary Statistics—Class Variables from the Output Window of the StatExplore node. Figure 3.8 shows the class (categorical) variable results. The results provide the count of each class (categorical) variable based upon the target value of Y or N. For example, there were 1400 observations that had a Bachelor Level for the Education variable and with no (N) Fraudulent_Claim and only 91 observations that had a Bachelor Level for the Education variable with yes (Y)Fraudulent_Claim.

Figure 3.9 displays the results for the interval variables. These results are the same results that were given in the Output Window of the SASExplore node shown in Fig. 3.7.

Results - Node: StatExplore Diagram: Data Preparation and Analysis
File Edit View Window

Class Variables

Data Role	Target	Target Level	Variable Name	Level	CODE	Frequency Count	Type	Percent Within	Level Index	Role	Label	Percent	Plot
TRAIN	Fraudulent_Claim	N	Education		1	11	C	0.234342	1	INPUT	Education	0.0022	1
TRAIN	Fraudulent_Claim	N	Education	Bachelor	0	1400	C	29.82531	2	INPUT	Education	0.279944	1
TRAIN	Fraudulent_Claim	Y	Education	Bachelor	1	91	C	29.64169	2	INPUT	Education	0.018196	1
TRAIN	Fraudulent_Claim	N	Education	College	2	1404	C	29.91052	3	INPUT	Education	0.280744	1
TRAIN	Fraudulent_Claim	Y	Education	College	0	88	C	28.6645	3	INPUT	Education	0.017596	1
TRAIN	Fraudulent_Claim	N	Education	Doctor	5	195	C	4.154239	4	INPUT	Education	0.038992	1
TRAIN	Fraudulent_Claim	Y	Education	Doctor	4	12	C	3.908795	4	INPUT	Education	0.0024	1
TRAIN	Fraudulent_Claim	N	Education	High School or Below	1	1318	C	28.0784	5	INPUT	Education	0.263547	1
TRAIN	Fraudulent_Claim	Y	Education	High School or Below	3	87	C	28.33876	5	INPUT	Education	0.017397	1
TRAIN	Fraudulent_Claim	N	Education	Master	3	366	C	7.797188	6	INPUT	Education	0.073185	1
TRAIN	Fraudulent_Claim	Y	Education	Master	2	29	C	9.446254	6	INPUT	Education	0.005799	1
TRAIN	Fraudulent_Claim	N	Employment_Status	Disabled	3	201	C	4.282062	1	INPUT	Employmen..	0.040192	1
TRAIN	Fraudulent_Claim	Y	Employment_Status	Disabled	3	12	C	3.908795	1	INPUT	Employmen..	0.0024	1
TRAIN	Fraudulent_Claim	N	Employment_Status	Employed	0	2952	C	62.88879	2	INPUT	Employmen..	0.590262	1
TRAIN	Fraudulent_Claim	Y	Employment_Status	Employed	0	174	C	56.67752	2	INPUT	Employmen..	0.034793	1
TRAIN	Fraudulent_Claim	N	Employment_Status	Medical Leave	2	219	C	4.66553	3	INPUT	Employmen..	0.043791	1
TRAIN	Fraudulent_Claim	Y	Employment_Status	Medical Leave	1	15	C	4.885993	3	INPUT	Employmen..	0.002999	1
TRAIN	Fraudulent_Claim	N	Employment_Status	Retired	4	151	C	3.216873	4	INPUT	Employmen..	0.030194	1
TRAIN	Fraudulent_Claim	Y	Employment_Status	Retired	4	7	C	2.28013	4	INPUT	Employmen..	0.0014	1
TRAIN	Fraudulent_Claim	N	Employment_Status	Unemployed	1	1171	C	24.94674	5	INPUT	Employmen..	0.234153	1
TRAIN	Fraudulent_Claim	Y	Employment_Status	Unemployed	2	99	C	32.24756	5	INPUT	Employmen..	0.019796	1
TRAIN	Fraudulent_Claim	N	Gender	F	0	2408	C	51.29953	1	INPUT	Gender	0.481504	0
TRAIN	Fraudulent_Claim	Y	Gender	F	1	62	C	20.19544	1	INPUT	Gender	0.012398	0
TRAIN	Fraudulent_Claim	N	Gender	M	1	2286	C	48.70047	2	INPUT	Gender	0.457109	0
TRAIN	Fraudulent_Claim	Y	Gender	M	0	245	C	79.80456	2	INPUT	Gender	0.04899	0
TRAIN	Fraudulent_Claim	N	Marital_Status	Divorced	2	713	C	15.1896	1	INPUT	Marital_Stat..	0.142571	0

Fig. 3.8 StatExplore node—class variable results

Results - Node: StatExplore Diagram: Data Preparation and Analysis
File Edit View Window

Interval Variables

Data Role	Target	Target Level	Variable	Median	Missing	Non Missing	Minimum	Maximum	Mean	Standard Deviation	Skewness	Kurtosis	Role	Label	Scaled Mean Deviation	Maximum Deviation	Level Id
TRAIN	Fraudulent_Claim	N	Income	34990	0	4694	0	15967801	41798.41	227890.4	68.48667	4790.542	INPUT	Income	0.011812	0.180605	1
TRAIN	Fraudulent_Claim	Y	Income	29735	0	307	0	98701	33849.56	30482.01	0.417743	-1.02688	INPUT	Income	-0.18061	0.180605	2
TRAIN	Fraudulent_Claim	N	Claim_Amount	581.0631	0	4694	189.8684	7422.852	792.7236	655.9633	2.922369	12.62942	INPUT	Claim_Amo..	0.006297	0.096276	1
TRAIN	Fraudulent_Claim	Y	Claim_Amount	534.5681	0	307	200.4351	3345.179	711.9203	577.7544	2.208033	4.72192	INPUT	Claim_Amo..	-0.09628	0.096276	2

Fig. 3.9 StatExplore node—interval variable results

3.4 The Shape of the Distribution

3.4.1 Skewness

The shape of a distribution describes how the data is distributed. The two common measures that describe the shape are skewness and kurtosis. The skewness and kurtosis values for the input variables are given in the Output Window of the StatExplore node in the interval variable statistics table (Fig. 3.5).

Skewness measures the extent to which input variables are not symmetrical. It measures the relative size of the two tails. A distribution can be left-skewed, symmetric, or right-skewed. In a left-skewed distribution, the mean is less than the median, and the skewness value is negative. Notice in Fig. 3.10 that the peak is on the right for a left-skewed distribution. For a normal, symmetrical distribution, the mean and the median are equal, and the skewness value is zero. For a right-skewed distribution, the mean is greater than the median, the peak is on the left, and the skewness value is positive. The formula to calculate the skewness is given in Fig. 3.11. If the skewness is greater than 2 or less than 2, the skewness is substantial and the distribution is not symmetrical (West et al. 1995; Trochim and Donnell 2006).

Fig. 3.10 Skewness

$$\frac{n}{(n-1)(n-2)}\sum_{i=1}^{n}\left(\frac{x_i-\bar{x}}{s}\right)$$

Where n is the sample size,
x$_i$ is the ith value of the variable,
$\bar{x}$ is the sample average, and
s is the sample standard deviation

Fig. 3.11 Skewness Formula

3.4.2 *Kurtosis*

Kurtosis measures how peaked the curve of the distribution is. In other words, how sharply the curve rises approaching the center of the distribution. It measures the amount of probability in the tails. The shape can be mesokurtic, leptokurtic, or platykurtic (Fig. 3.12). The kurtosis value is calculated using the formula given in Fig. 3.13. A mesokurtic distribution is a normal bell-shaped distribution, and the kurtosis value is equal to 3. A leptokurtic has a sharper peak than a bell shape, and it has heavier tails than a normal distribution. A leptokurtic distribution has a kurtosis value of greater than 3. A platykurtic distribution is flatter and has smaller tails. The kurtosis value is less than 3. An absolute kurtosis value >7.1 is considered a substantial departure from normality (West et al. 1995). Many statistical software packages provide a statistic titled "excess kurtosis" which is calculated by subtracting 3 from the kurtosis value. The excess kurtosis value of zero indicates a perfectly symmetrical normal distribution.

When performing predictive modeling, the distribution of the variables should be evaluated. If it is highly skewed, a small percentage of the data points (i.e., those lying in the tails of the distribution) may have a great deal of influence on the predictive model. In addition, the number of variables available to predict the target variable can vary greatly. To counteract the impact of skewness or kurtosis, the variable can be transformed. There are three strategies to overcome these problems:

1. Use a transformation function, to stabilize the variance.
2. Use a binning transformation, which divides the variable values into groups to appropriately weight each range.
3. Use both a transformation function and a binning transformation. This transformation can result in a better-fit model.

Fig. 3.12 Kurtosis taken from https://www.bogleheads.org/wiki/Excess_kurtosis

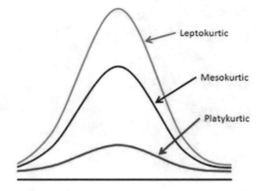

$$\frac{n(n+1)}{(n-1)(n-2)(n-3)}\sum_{i=1}^{n}\left(\frac{(x_i-\bar{X})}{S}\right)^4 - \frac{3(n-1)^2}{(n-2)(n-3)}$$

Where n is the sample size,

x$_i$ is the ith value of the variable,

$\bar{x}$ is the sample average, and

s is the sample standard deviation

Fig. 3.13 Kurtosis formula

Transformations are more effective when there is a relatively wide range of values as opposed to a relatively small range of values (e.g., total sales for companies can range significantly by the company's size). The logarithmic function is the most widely used transformation method to deal with skewed data. The log transformation is used to transform skewed data to follow an approximately normal distribution.

Returning to the automobile insurance claim fraud example, the Output Window from the StatExplore, shown earlier in Fig. 3.5, illustrated the skewness and kurtosis values for the claim_amount and income. For the variable claim_amount, the skewness value was 2.922 and kurtosis 12.62. Notice that the skewness and kurtosis values for income were $68.48667 and $4790.542, respectively. Income is highly right-skewed with a leptokurtic shape. Income should be transformed to provide a more accurate model. The Transform node can be used to modify the income variable. From the Modify tab, drag a Transform Variables node (Fig. 3.14) to the process flow diagram (Fig. 3.15), and connect it to the Filter node.

In the Train group of the Transform Properties window, select the ellipsis next to the Formulas property (Fig. 3.16).

Fig. 3.14 Transform
Variables node

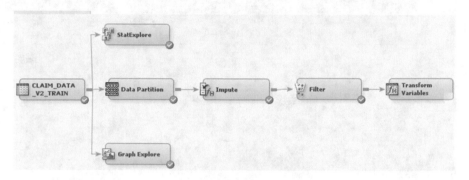

Fig. 3.15 Process flow diagram with Transform Variables node

Fig. 3.16 Transform
Variables properties

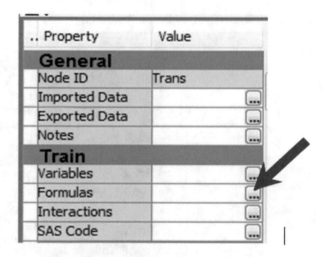

The Formulas window will open. In this window, you can view the distribution
of your variables. Select the various input variables to see the distributions. Notice
that Income is right-skewed (Fig. 3.17). The log transformation will be used to control for this skewness.

In the Train group, select the ellipsis button next to the Variables property
(Fig. 3.18). The Variable window will open. Change the method for the Income variable from Default to Log (Fig. 3.19).

In the Transform Properties window, set the default method for interval inputs to
none (Fig. 3.20). In this case, the interval variables do not require binning. Binning

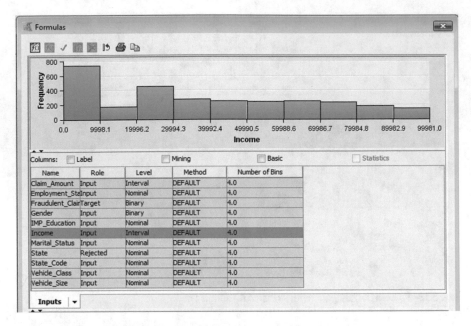

Fig. 3.17 Formulas window—income distribution

Fig. 3.18 Transform
Properties window

is a method to categorize continuous predictor variables. The decision on when to
use binning is not clear cut. But, there are a number of research papers that discuss
this topic (Fayyad and Irani 1992; Grace-Martin 2018). By binning there is a loss of
information and power and if not necessary should be avoided. Run the Transform
Variables node and then view the results (Fig. 3.21). The data resulting from the
Transform Variables node now contains a new variable, Log_ Income. The new
variable has the prefix of the transformation method that was used. In addition, the
original variable still exists but now has a role set to be rejected.

Fig. 3.19 Transform Variables window

Fig. 3.20 Transform
Properties window

3.5   Covariance and Correlation

It is important to investigate the relationship of the input variables to one another
and to the target variable. The input variables should be independent of one another.
If the input variables are too related (i.e., correlated) to one another, multicollinearity can occur which affects the accuracy of the model. Both the covariance and correlation describe how two variables are related. The *covariance* indicates whether

Fig. 3.21 Results of Transform node

$$COV(x,y) = \frac{\sum_{i=1}^{n}(x_i - \bar{x})(y_i - \bar{y})}{n-1}$$

Where n is the sample size,
x_i is the ith value of the variable x
$\bar{x}$ is the average of x,
v_i is the ith value of the variable y,
$\bar{y}$ is the average of y

Fig. 3.22 Covariance formula

two variables are positively or inversely related. A measure used to indicate the extent to which two random variables change in tandem is known as covariance. The formula for the covariance is given in Fig. 3.22.

The covariance is only concerned with the strength of the relationship; there is no causal effect implied. If the covariance between two variables, denoted as x (input variable) and y (target variable), is greater than zero, then the two variables tend to move in the same direction. That is, if x (input variable) increases, then y (target variable) will also tend to increase. Conversely, if the covariance is less than zero, x (input variable) and y (target variable) will tend to move in opposing or opposite directions. If the covariance of x (input variable) and y (target variable) are equal to zero, they are independent. A major weakness with covariance that it is not possible to determine the relative strength of the relationship from the size of the covariance. The covariance value is the product of the two variables and is not a standardized unit of measurement. So, it is impossible to measure the degree the variables move together.

The *correlation* does measure the relative strength of the relationship. It standardizes the measures so that two variables can be compared. The formula for the correlation (referred to as r) is given in Fig. 3.23. The correlation ranges from −1 to +1. A correlation value of −1 indicates an inverse correlation, +1 indicates a positive correlation, and 0 indicates no correlation exists. The correlation (r) is normally provided but, squaring the correlation, provides some additional value. The r-squared (r^2) or coefficient of determination is equal to the percent of the variation in the target variable that is explained by the input variable. R-squared ranges from 0% to 100%. For example, an r^2 of 0.25 indicates that 25% of the variation in target variable (y) is explained by the input variable (x). The relationship is often viewed using a scatter plot. A scatter plot is a chart where two different variables are plotted, one on each axis (Fig. 3.24). In regression, the input variable, x, is on the x-axis, and the target variable, y, is on the y-axis.

If there is a strong correlation between the input variables, there can be some multicollinearity, which may negatively affect the predictive model. This undermines the assumption of independent input variables. In SAS Enterprise Miner™,

$$r_{x,y} = \frac{\text{cov}(x,y)}{S_x S_y}$$ Where $S_x S_y$ are the sample standard deviation of x and y.

Fig. 3.23 Correlation formula

Fig. 3.24 Correlation

Fig. 3.25 Variable Cluster node

correlation information can be obtained with the Variable Clustering node. The Variable Cluster node is available from the Explore tab (Fig. 3.25).

To find the correlation results, click on the View menu, and select Model and then Variable Correlation. A colored matrix will appear (Fig. 3.26). In general, the darker the red color, the higher the correlation between the two intersecting variables, and the darker the blue color, the more negative the correlation. Scrolling over each of the boxes shows the correlation values. Clicking on the Table icon (Fig. 3.27) shows the numeric correlation values (Fig. 3.28). Income and Outstanding_Balance are not correlated, and the absolute value of correlation value is 0.00394. There is no collinearity to warrant concern.

3.6   Variable Reduction

Reducing the number of variables can reduce multicollinearity, redundancy, and irrelevancy and improve the processing time of the model. Two methods for variable reduction include variable clustering and principal component analysis.

Fig. 3.26 Variable correlation matrix

Fig. 3.27 Table icon

Table: Variable Correlation			
Variable	Variable	Correlation ▲	
Claim_Amount	Income		-0.00394
Income	Claim_Amount		-0.00394
Claim_Amount	Claim_Amount		1
Income	Income		1

Fig. 3.28 Correlation results

3.6.1   Variable Clustering

Variable clustering identifies the correlations and covariances between the input variables and creates groups or clusters of similar variables. Clustering attempts to reduce the correlation within the groups. A few representative variables that are fairly independent of one another can then be selected from each cluster. The representative variables are used as input variables, and the other input variables are rejected.

To determine the clusters, the Euclidean distance is calculated. The *Euclidean distance* is the straight-line distance between two points in Euclidean space. The Pythagorean theorem is used to calculate the distance between the two points. The theorem states that for a right triangle the square of the length of the hypotenuse (the line opposite the right angle) is equal to the sum of the squares of the lengths of the other two sides (Fig. 3.29). Figure 3.30 shows how the theorem is applied for a two-dimensional space.

For points in three-dimensional space, (X_1, Y_1, Z_1) and (X_2, Y_2, Z_2), the Euclidean distance between them is $\sqrt{(X_2 - X_1)^2 + (Y_2 - Y_1)^2 + (Z_2 - Z_1)^2}$.

Figure 3.31 works through an example of variable clustering using just four data points to explain how the process works using the Euclidean distance formula. The centroid is the average of the points in the cluster. Increasing the number of variables increases the amount of work to calculate the clusters and becomes more time-consuming.

Variable Clustering Using Euclidean Distance Example
Consider the following scatter chart of four data points (Fig. 3.31). At the start, each point is its own cluster and is the centroid of that cluster. The general rule for combining points or clusters is to combine the centroid points that have the smallest distance from one another.

The next step is to determine the two pairs of data points that are the closest to one another. Calculation of the Euclidean distance for all possible combinations is shown in Table 3.1.

The two points closest to each other are points with the lowest Euclidean distance, points (1,5) and (2,4). Points 1 and 2 are combined to form cluster 1. The centroid for the cluster is the average of the two points, equal to (1.5, 4.5) (Fig. 3.32).

Once again, the distance between the points are calculated but this time using the centroid value of cluster 1 (Table 3.2).

Points (4,8) and (5,6) are the closest, so they are combined to form cluster 2. The centroid for cluster 2 is (4.5, 7) (Fig. 3.33).

At this point, we have two clusters. It does not make sense to combine the two clusters. How do you know when to stop combining clusters? Depending on the situation, there may be a reason to pre-define the number of clusters or to set the maximum distance between the centroid and the points.

In SAS Enterprise Miner™, the Variable Clustering node can be used to create the different clusters and select the representative variables from each cluster. The cluster variable node can be placed after the Transform node. The properties of the

Fig. 3.29 Illustration of Pythagorean theorem

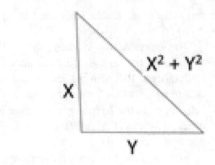

Fig. 3.30 Euclidean distance in two-dimensional space

Fig. 3.31 Scatter plot

Variable Clustering node are shown in Fig. 3.34. If the data has more than 30 variables, the Keep Hierarchies property should be set to yes, and the Two-Stage Clustering property should be set to yes (Brown 2017). The Variable Selection property determines the method for identifying the variables that will be passed to the subsequent node. The default method, cluster component, passes a linear

Table 3.1 Euclidean distance iteration 1

Start (x,y)	End (x,y)	Euclidean distance
1,5	2,4	$1.41 = \sqrt{(2-1)^2 + (4-5)^2}$
1,5	4,8	4.24
1,5	5,6	4.12
2,4	4,8	4.47
2,4	5,6	3.61
4,8	5,6	2.24

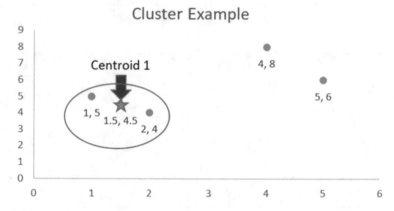

Fig. 3.32 Scatter plot with one cluster

Table 3.2 Euclidean distance table—iteration 2

Start (x,y)	End (x,y)	Euclidean distance
(1.5, 4.5)	4,8	4.30
(1.5, 4.5)	5,6	3.81
4, 8	5,6	2.24

Fig. 3.33 Scatter plot with two clusters

Fig. 3.34 Cluster variable
node properties

Train		
Variables		[...]
Clustering Source	Correlation	
Keeps Hierarchies	Yes	
Includes Class Variables	No	
Two Stage Clustering	Auto	
⊟ Stopping Criteria		
⋮ Maximum Clusters	.	
⋮ Maximum Eigenvalue	.	
⋮ Variation Proportion	0.0	
Print Option	Short	
Suppress Sampling Warning	No	

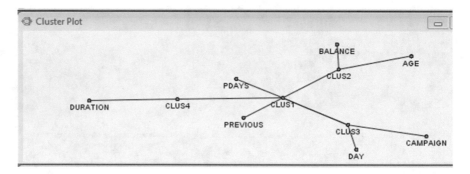

Fig. 3.35 Bank marketing cluster plot

combination of the variables from each cluster. The alternative method, best variables, selects the best variables in each cluster that have the minimum r^2 ratio value. Clustering is useful if the data set has fewer than 100 variables and fewer than 100,000 observations. If there are more than 100 variables, two-stage variable clustering should be used (Brown 2017). To illustrate the different variable reduction techniques, an example based on bank marketing data is given. The resulting cluster plot provides a visual representation of the different clusters (Fig. 3.35).

The resulting dendrogram is provided in Fig. 3.36. The dendrogram shows the hierarchical relationship between the objects. The dendrogram provides a visual representation of the relatedness of variables within a cluster. The height of the horizontal lines indicates the degree of difference between branches. The longer the line, the greater the difference. The dendrogram is read left to right. The relationships between the variables Pdays and Previous, Campaign and Day, and Age and Balance are the most similar and are first joined together. Duration is connected with this cluster. This indicates that every variable within that cluster is more like each other than to any variable or cluster that joins at a higher level.

The Variable Selection Table (Fig. 3.37) shows the variables within each cluster and the squared correlations (r^2) with its own cluster component and the r^2 of the next highest squared correlation with a cluster component. The within value should

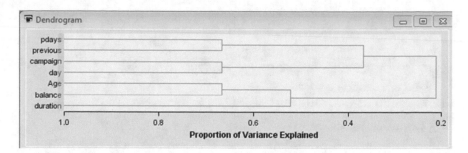

Fig. 3.36 Bank marketing dendrogram

Cluster	Variable	R-Square With Own Cluster Component	Next Closest Cluster	R-Square with Next Cluster Component	Type	Label	1-R2 Ratio	Variable Selected
CLUS1	CLUS1	1	CLUS3	0.011469	ClusterComp	Cluster 1	0	YES
CLUS1	PREVIOUS	0.703759	CLUS3	0.002767	Variable		0.297063	NO
CLUS1	PDAYS	0.703759	CLUS3	0.016148	Variable		0.301103	NO
CLUS2	CLUS2	1	CLUS1	0.001557	ClusterComp	Cluster 2	0	YES
CLUS2	AGE	0.555336	CLUS1	.0004614	Variable		0.444869	NO
CLUS2	BALANCE	0.555336	CLUS1	0.001394	Variable		0.445284	NO
CLUS3	CLUS3	1	CLUS1	0.011469	ClusterComp	Cluster 3	0	YES
CLUS3	DAY	0.56874	CLUS1	0.004174	Variable		0.433068	NO
CLUS3	CAMPAIGN	0.56874	CLUS1	0.009393	Variable		0.435349	NO
CLUS4	CLUS4	1	CLUS1	0.001309	ClusterComp	Cluster 4	0	YES
CLUS4	DURATION	1	CLUS1	0.001309	Variable		0	NO

Fig. 3.37 Variable clustering—Variable Selection Table

Cluster	Frequency Count	Percent of Total Frequency
CLUS1	38	65.51724
CLUS2	5	8.62069
CLUS3	9	15.51724
CLUS4	3	5.172414
CLUS5	3	5.172414

Fig. 3.38 Variable Frequency Table

be higher than the r^2 with any other cluster unless an iteration limit has been exceeded or the CENTROID option has been used. The larger the r^2 is, the better. The column labeled "Next Closest" contains the next highest squared correlation of the variable with a cluster component. This value is low if the clusters are well separated. The $1- R^2$ ratio displays the ratio of $(1- R^2_{\text{with own}})/(1- R^2_{\text{next}})$ for each variable. Small values of this ratio indicate good clustering. The Frequency Table (Fig. 3.38) provides the frequency or the number of observations put in each cluster and the corresponding frequency percentage of the total frequency.

Figure 3.39 shows two snippets of the output created by the Variable Clustering node. Notice that cluster 1 is split because its second Eigenvalue is greater than 1. The Eigenvalue for a given cluster measures the variance in all the variables, which

```
33
34                        Cluster Summary for 1 Cluster
35
36                         Cluster   Variation   Proportion      Second
37    Cluster   Members   Variation  Explained   Explained     Eigenvalue
38    --------------------------------------------------------------------
39        1        7          7       1.47359     0.2105         1.1110
40
41    Total variation explained = 1.47359 Proportion = 0.2105
42
43    Cluster 1 will be split because it has the largest second eigenvalue, 1.111008, which is greater than the MAXEIGEN=1 value.
44
45
46    Clustering algorithm converged.
47
48
49                        Cluster Summary for 2 Clusters
50
51                         Cluster   Variation   Proportion      Second
52    Cluster   Members   Variation  Explained   Explained     Eigenvalue
53    --------------------------------------------------------------------
54        1        4          4       1.465425    0.3664         1.0069
55        2        3          3       1.112688    0.3709         1.0002
56
57    Total variation explained = 2.578114 Proportion = 0.3683
```

```
110    Clustering algorithm converged.
111
112
113                        Cluster Summary for 4 Clusters
114
115                         Cluster   Variation   Proportion      Second
116    Cluster   Members   Variation  Explained   Explained    Eigenvalue
117    --------------------------------------------------------------------
118       1         2          2       1.407518    0.7038        0.5925
119       2         2          2       1.110673    0.5553        0.8893
120       3         2          2       1.13748     0.5687        0.8625
121       4         1          1       1           1.0000
122
123    Total variation explained = 4.655671 Proportion = 0.6651
124
125
126                             R-squared with
127    4 Clusters              --------------------
128                             Own        Next      1-R**2
129    Cluster     Variable    Cluster    Closest    Ratio
130    --------------------------------------------------------------------
131    Cluster 1   pdays       0.7038     0.0161     0.3011
132                previous    0.7038     0.0028     0.2971
133    --------------------------------------------------------------------
134    Cluster 2   Age         0.5553     0.0005     0.4449
135                balance     0.5553     0.0014     0.4453
136    --------------------------------------------------------------------
137    Cluster 3   campaign    0.5687     0.0094     0.4353
138                day         0.5687     0.0042     0.4331
139    --------------------------------------------------------------------
140    Cluster 4   duration    1.0000     0.0013     0.0000
141
142    No cluster meets the criterion for splitting.
```

Fig. 3.39   Variable Clustering node output

is accounted for by that cluster. The clusters will continue to be split based upon the largest second Eigenvalue. Splitting will be completed when all the Eigenvalues are less than or equal to 1 or until it is not possible for the cluster to be split any further (assuming the maximum number of splits was not set). Notice at the creation of the fourth cluster, the second Eigenvalue is not given+ this is because cluster 4 could not be split any further. The default maximum Eigenvalue is 1. However, this can be changed in the corresponding Variable Clustering ode property. The output also displays the variation explained by the clusters, which is based only on the variables in the cluster. The *Proportion Explained* is calculated by dividing the *Cluster Explained* by the *Cluster Variation* (Fig. 3.40).

3.6.2   Principal Component Analysis

Principal component analysis (PCA) is another variable reduction strategy. It is used when there are several redundant variables or variables that are correlated with one another and may be measuring the same construct. Principal component analysis mathematically manipulates the input variables and develops a smaller number of artificial variables (called principal components). These components are then used in the subsequent nodes. The first principal component is created so that it

```
125
126                                   R-squared with
127    4 Clusters                     ------------------
128                            Own       Next      1-R**2
129    Cluster    Variable    Cluster   Closest    Ratio
130    -----------------------------------------------------
131    Cluster 1  pdays       0.7038    0.0161    0.3011
132               previous    0.7038    0.0028    0.2971
133    -----------------------------------------------------
134    Cluster 2  Age         0.5553    0.0005    0.4449
135               balance     0.5553    0.0014    0.4453
136    -----------------------------------------------------
137    Cluster 3  campaign    0.5687    0.0094    0.4353
138               day         0.5687    0.0042    0.4331
139    -----------------------------------------------------
140    Cluster 4  duration    1.0000    0.0013    0.0000
141
142    No cluster meets the criterion for splitting.
143
144
145               Total     Proportion    Minimum     Maximum                 Maximum
146               Variation    of        Proportion   Second      Minimum     1-R**2
147    Number    Explained  Variation    Explained   Eigenvalue  R-squared    Ratio
148    of          by       Explained     by a         in a       for a      for a
149    Clusters  Clusters   by Clusters   Cluster     Cluster    Variable    Variable
150    -----------------------------------------------------------------------------------
151       1      1.473590     0.2105      0.2105      1.111008    0.0047
152       2      2.578114     0.3683      0.3664      1.086885    0.0197      0.9808
153       3      3.657687     0.5225      0.3709      1.000191    0.0197      0.9815
154       4      4.655671     0.6651      0.5553      0.889327    0.5553      0.4453
155
```

Fig. 3.40 Variable Clustering node output—Proportion Explained captures as much of the variation in the input variables as possible. The second principal component accounts for as much of the remaining variation as possible and so forth with each component maximizing the remaining variation. The number of principal components can be based upon the proportion of variance explained, the scree plot (Eigenvalue plot), or an Eigenvalue greater than 1. One major drawback of using PCA is that it is very difficult to interpret the principal components and it is hard to determine how many components are needed (Brown 2017). A scree plot is a line segment graph that displays a decreasing function that shows the total variance explained by each principal component.

The Principal Components node is located on the Modify tab (Fig. 3.41). The corresponding node properties are shown in Fig. 3.42. The maximum number of principal components is defined by the maximum property (located within the maximum number cutoff section). Figure 3.43 shows the Eigenvalue or scree plot. The plot shows the Eigenvalues on the y-axis and the number of principal components on the x-axis. It will always be a downward curve; the point where the slope of the curve flattens indicates the number of principal components needed. The plot for the bank marketing data example suggests the use of 17 principal components.

The output generated by the principal components (Fig. 3.44) shows the Eigenvalues and the proportion of variation explained by the principal components. The Difference column values are calculated by taking the Eigenvalue of the component subtracted by

Fig. 3.41 Principal Components icon

Fig. 3.42 Principal Components node properties

Train	
Variables	...
Eigenvalue Source	Correlation
Interactive Selection	...
Print Eigenvalue Source	No
Score	
Princomp Prefix	PC
⊟Eigenvalue Cutoff	
├Cumulative	0.99
└Increment	0.001
⊟Max Number Cutoff	
├Apply Maximum Number	Yes
└Maximum Number	20
Reject Original Input Variab	Yes
Hide Rejected Variables	Yes

Fig. 3.43 Eigenvalue plot

```
404     The DMNEURL Procedure
405
406                   Eigenvalues of Correlation Matrix
407
408             Eigenvalue    Difference    Proportion    Cumulative
409
410      1      17.1261160    6.9577215       0.0464        0.0464
411      2      10.1683944    4.2296284       0.0276        0.0740
412      3       5.9387660    1.1282067       0.0161        0.0901
413      4       4.8105593    0.7846555       0.0130        0.1031
414      5       4.0259038    0.2743280       0.0109        0.1140
415      6       3.7515758    0.1470459       0.0102        0.1242
416      7       3.6045299    0.2169130       0.0098        0.1339
417      8       3.3876169    0.2830390       0.0092        0.1431
418      9       3.1045779    0.0230903       0.0084        0.1515
419     10       3.0814876    0.2189337       0.0084        0.1599
420     11       2.8625539    0.1534329       0.0078        0.1676
421     12       2.7091210    0.0646511       0.0073        0.1750
422     13       2.6444699    0.0463002       0.0072        0.1822
423     14       2.5981697    0.0232850       0.0070        0.1892
424     15       2.5748846    0.0745459       0.0070        0.1962
425     16       2.5003387    0.0169764       0.0068        0.2030
426     17       2.4833623    0.0381611       0.0067        0.2097
427     18       2.4452012    0.0065952       0.0066        0.2163
428     19       2.4386060    0.0196512       0.0066        0.2229
429     20       2.4189547    0.0458049       0.0066        0.2295
```

Fig. 3.44 Principal component output

the next component's Eigenvalue. The proportion is calculated by the Eigenvalue divided by the total Eigenvalue (the sum of the column). Looking at the Cumulative column, 4.64% of the variation is explained by this first component, and 7.4% of the variation is explained by the first two components. Notice around the 17th component the proportion value does not improve significantly.

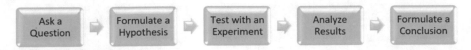

Fig. 3.45 Scientific method

Fig. 3.46 Hypothesis
example

> H_o: Average Wait Time is 15 min. $(\mu = 15)$
> H_A: Average Wait Time is not 15 min. $(\mu \neq 15)$

3.7   Hypothesis Testing

A *hypothesis* is a supposition or observation regarding the results of sample data. It is the second step in the scientific method (Fig. 3.45). In hypothesis testing, information is known about the sample. The purpose of hypothesis testing is to see if the hypothesis can be extended to describe the population. For example, a sample of hospital wait times is 15 min. The hypothesis may state that average wait time for the hospital population is 15 min. The null hypothesis states the supposition to be tested. The alternative hypothesis is the opposite of null hypothesis (Fig. 3.46).

Begin with the assumption that the null hypothesis is true. This is analogous to the rule of law that assumes innocence until proven guilty in the US court system. It represents the belief in a situation. It is always stated with the logical operators, equals, less than or equal to, greater than, or greater than or equal to. The hypothesis may or may not be rejected. The null hypothesis is never accepted, and the result is either fail to reject or reject the null.

In hypothesis testing there are two types of errors that can occur. A type I and type II error. A type I occurs when you reject a true null hypothesis. This is like finding an innocent person guilty. It's a false alarm. The probability of a type I error is referred to as alpha, α, and it's called the level of significance of the test. A type II error, called beta, is the failure to reject a false null hypothesis. It's equivalent to letting a guilty person go. A type II error represents a missed opportunity. The power of the test is the probability of rejecting a false null hypothesis. It is equivalent to finding a guilty person guilty. The power of the test is equal to 1 minus beta. The steps to hypothesis testing are as follows:

1. Specify the null hypothesis.
2. Specify the alternative hypothesis.
3. Select the appropriate test statistic and the significance level (alpha, α).
4. Calculate the test statistic and corresponding *p*-value.
5. Draw a conclusion.

The appropriate test statistic is based upon the hypothesis. For example, if the test is on the hypothesis of a mean with an unknown population standard deviation and a normal distribution, the t-statistic is used; if the population standard deviation is known and the data is normally distributed, a z-statistic is used. The level of significance is the value of α, and it is the probability of rejecting the null when it is true. This level is set at the start of the analysis and is based on the severity of incorrectly rejecting the null. Usually α is set to 0.05% or 5%, but other levels commonly used are 0.01 and 0.10.

Once the appropriate test statistic is calculated, the p-value can be determined. The p-value is the smallest significance level that leads to rejection of the null hypothesis. Alternatively, the p-value is the probability that we would observe a more extreme statistic than we did if the null hypothesis were true. If the p-value is small, meaning the p-value is less than alpha, then we reject the null hypothesis. The p-value is the evidence against a null hypothesis. The smaller the p-value, the stronger the evidence that you should reject the null hypothesis. If the p-value is less than (or equal to) α, reject the null hypothesis in favor of the alternative hypothesis. If the p-value is greater than α, do not reject the null hypothesis. Figure 3.44 illustrates an example hypothesis test on the mean with the population standard deviation unknown.

Hypothesis Test on Mean Example
A random sample of 25 customers is taken, and the resulting sample mean was determined to be 16 min with a sample standard deviation of 3 min. You want to test the hypothesis at $\alpha = 0.05$. Assume the population distribution is normal.

Since the population standard deviation is not known, the t-test is the appropriate statistic to use. The formula is shown in Fig. 3.47.

Computing the area in the tail region, at $x = 1.667$, you get an area or p-value of 0.10858. Since the p-value is greater than α, you do not reject the null hypothesis and conclude there is insufficient evidence that the population mean time is different from 15 min. A graphical representation of the problem is shown in Fig. 3.48.

3.8 Analysis of Variance (ANOVA)

ANOVA is used to test the difference among the means of input variables. It can be broken into two categories: one-way ANOVA and two-way ANOVA. In one-way ANOVA, the difference among the means of three or more inputs is evaluated to see if any of them are equal. Two-way ANOVA examines the effect of two inputs on the target variable. In other words, it examines the interaction between two input variables on the target variable.

The null hypothesis for one-way ANOVA is that the means for the different input variables are equal (H_o: $\mu_1 = \mu_2 = \mu_3 = \mu_4 = \dots$). The alternative hypothesis is not all input variable means are equal. There is a factor effect. At least one input mean is different. The test statistic is an F statistic equal to the mean squared among the

Fig. 3.47 T-statistic formula

$$t_{\text{statistic}} = \frac{X-\mu}{S/\sqrt{n}} = \frac{16-15}{3/\sqrt{25}} = 1.667, \text{ the critical } t_{\alpha/2} = .025$$

Fig. 3.48 Hypothesis testing illustration

groups divided by the mean squared within groups. Most statistical software, including SAS Enterprise Miner™, will provide the F statistic and the associated p-value. The null hypothesis is rejected if the F statistic value is greater than the F critical value or alternatively if the p-value is less than the level of significance. If the null hypothesis is rejected, then all the input means are not equal, and at least one of them is different.

The null hypothesis for two-way ANOVA is the interaction effect of two input variables being equal to zero. The alternative hypothesis is the interaction effect of two input variables that is not equal to zero. This means there is an interaction effect. The test statistic is an F statistic equal to the mean square interaction divided by the mean squared error. Again, most statistical software, including SAS Enterprise Miner™, will provide the F statistic and the associated p-value. The null hypothesis is rejected if the F statistic value is greater than the F critical value or alternatively if the p-value is less than the level of significance. If the null hypothesis is rejected, then the interaction of the two inputs is not zero.

3.9   Chi-Square

The *Chi-square test* is used to determine if two categorical (class) variables are independent. The null hypothesis is that the two categorical variables are independent. The alternative is that they are not independent. The Chi-square test statistic approximately follows a Chi-square distribution with $(r-1)*(c-1)$ degrees of freedom, where r is the number of levels for one categorical variable and c is the number of levels for the other categorical variable. The Chi-square test statistic is shown in Fig. 3.49.

If the Chi-square statistic is greater than the critical Chi-square, then the null hypothesis is rejected. The critical Chi-square value is based upon the level of significance. An example of a problem that tests the independence of two categorical variables, gender, and education is given below.

$$\chi^2 = \sum (O - E)^2 / E$$
Where O is the observed frequency,
E is the expected frequency, if the null hypothesis were true

Fig. 3.49 Chi-square test statistic

Chi-Square Test Example
We want to know if gender and education are related at a 5% level of significance. A random sample of 400 people is surveyed.

H_o: Gender and education are independent.
H_A: Gender and education are not independent.

A summarized observed frequency table along with a table of the expected values, if the null hypothesis is true, is given in Table 3.3.

3.10 Fit Statistics

To measure the accuracy of a data model, fit statistics such as the misclassification rate, the receiver operating characteristic (ROC), and the average squared error are often evaluated. A fit statistic is a measure used to compare multiple models. A *misclassification* occurs when the predicted target value is not equal to the actual target value (indicating either a type I or type II error). A low misclassification rate is more desirable and indicates a better-fit model.

The *receiver operating characteristic curve* or *ROC curve* is a plot of the true positive rate against the false positive rate at various possible outcomes (Fig. 3.44). The true positive rate or sensitivity is the measure of the proportion of actual positives that were correctly identified. For example, the true positive rate would be the percentage of fraudulent claims that were correctly identified as fraudulent. This is equivalent to the power of the test in hypothesis testing. The false positive rate or 1-specificity is the measure of the proportion of positives that were incorrectly identified. This is equivalent to a type I error in hypothesis testing. For example, the false-positive rate would be the percentage of valid claims that were incorrectly identified as fraudulent.

The closer a ROC curve follows the upper left corner, the more accurate the model. On the other hand, the closer the plots are to the diagonal, the less accurate the model is. The C-statistic which is equal to the area under the ROC curve can be used as a criterion to measure the fit of a model. The greater the area under the curve, the more accurate the model. A perfect model would have a C-statistic equal to 1. A value of 0.8 indicates a strong model. The diagonal line represents a random

Table 3.3 Chi-square frequency

Observed frequencies

	Education				
	High school	Bachelor	Master	Doctorate	Total
Male	42	80	54	47	223
Female	60	58	46	13	177
Total	102	138	100	60	400

Expected frequencies

	Education				
Opinion	High school	Bachelor	Master	Doctorate	Total
Male	56.865	76.935	55.75	33.45	223
Female	45.135	61.065	44.25	26.55	177
Total	102	138	100	60	400

$\chi^2 = (42-56.865)^2/56.865 + \cdots + (12-26.55)^2/26.55 = 21.5859$

The critical Chi-square value at the 0.05 level of significance and $(2-1) \times (3.1) = 3$ degrees of freedom = 7.8145

Since $21.5859 > 7.815$, we reject the null and conclude that the education level depends on gender at a 5% level of significance.

chance and has a value of 0.5. This model would not be useful, it is correct only 50% of the time, and the results would be the same as just flipping a coin (Fig. 3.50).

The *average squared error* is the third measure for evaluating the fit of a model. The average squared differences are the average over all observations of the squared differences between the target value and the estimate of the target value. The average squared error is related to the amount of bias in a model. A model with a lower average squared error is less bias.

3.11   Stochastic Models

A *stochastic model* is a mathematical method that accounts for random variation in one or more inputs. The random variation is usually based on past observations. It is better described by comparing a stochastic model to a deterministic one. A deterministic model generally only has one set of output values based upon the parameter values and the initial conditions. A stochastic model will use the same set of parameter values and initial conditions but adds some random variation to the model resulting in a set of different outputs. The random variation is generated from observations of historical data over a period of time. The set of outputs are usually generated through the use of many simulations with random variations in the inputs. One well-known stochastic model is the Monte Carlo method. The steps to the Monte Carlo method are:

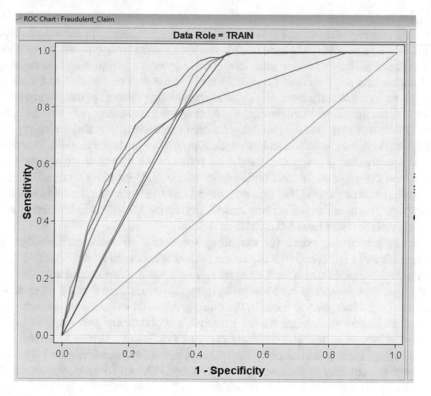

Fig. 3.50 ROC curve

1. Define a domain of possible inputs. The domain describes the distribution of the inputs. For example, the inputs may be normally or uniformly distributed with specific means and standard deviations.
2. Generate inputs randomly from the probability distribution.
3. Calculate the results based on the inputs.
4. Repeat steps 2 and 3 to generate a set of results.
5. Aggregate the results.

3.12   Chapter Summary

Descriptive analytics mines and prepares the data for creating predictive analytics models. It is an important but time-consuming task. It is estimated that approximately 60% of data scientists spend most of their time on data organization and cleaning (Press 2016). Prior to applying descriptive analytics, the business problem and hypotheses must be determined. Then the data is collected. Next we apply descriptive analytics. Chapter 2 described the data types and roles, identified outliers and missing values, and how to handle these situations. In SAS Enterprise Miner™, a project file, diagram, library, and data source were created. Imputation, filter, and replacement nodes were introduced as well as partitioning into training, validation, and test data sets. Chapter 3 extended the data preparation and analysis step to applying descriptive analytics. When evaluating data, a hypothesis about the data is often proposed. For example, when comparing two categorical variables, the hypothesis would be that the two variables are independent. The alternative hypothesis would be they were not independent. Descriptive analytics initially examines these hypotheses by describing the data in terms of the central tendency, the variance, and the shape of the distribution of its variables. Depending on the shape and characteristics of the data, the data may need to be transformed to result in a better-fit model. The transformation node in SAS Enterprise Miner™ can be used to normalize variables that have a skewed distribution.

When performing predictive modeling, too many variables, collinearity, and redundancy can affect both the fit and the performance of the model. Variable clustering and principal component analysis can help reduce these problems. Variable clustering groups a set of variables into groups, called clusters that are more correlated to each other than to those in the other groups. To calculate the closest variables, the Euclidean distance is used. Principal component analysis mathematically alters the input variables and develops a smaller number of artificial variables.

When comparing the means of interval inputs, one-way ANOVA can be used to evaluate the equality of the means. Two-way ANOVA can be used to evaluate the effect of an interaction between two input variables on the target variable. When performing a test on categorical variables, the Chi-square test statistic is used.

Once the data is prepared and the model is generated, there are several statistics that can be used to evaluate the fit of the model. The ROC chart plots the true positive proportion and the false-positive proportion at various outcomes. The C-statistic or the area of the curve indicates the accuracy of the model. A value closer to 1 indicates a strong model. The misclassification rate and the average squared error may also be used to evaluate the fit of the model. In the subsequent chapters, predictive analytics techniques will be described. Fit statistics provide a basis for comparing multiple predictive analytics models.

Discussion Questions
1. Discuss the three measures of tendency.
2. Discuss the measures of variance.
3. Describe skewness and kurtosis. What actions should be taken if the data is skewed?
4. Describe methods for variable reduction.
5. What statistics can be used to evaluate the fit of the model?
6. Describe when it is appropriate to use ANOVA and Chi-square analysis.

References

Brown I (2017) Deeper dive: variable selection routines in SAS Enterprise Miner. Presented at the Analytics & Innovation Practice, SAS UK, on March 23, 2017. Available at https://www.sas.com/content/dam/SAS/en_gb/doc/presentations/user-groups/deeper-dive-variable-selection-routines-in-sas.pdf. Accessed on 27 Sept 2018

Fayyad U, Irani K (1992) On the handling of continuous-valued attributes in decision tree generation. Mach Learn 8(1):87–102

Grace-Martin K (2018) 3 Situations when it makes sense to categorize a continuous predictor in a regression model. The Analysis Factor. Taken from: https://www.theanalysisfactor.com/3-situations-when-it-makes-sense-to-categorize-a-continuous-predictor-in-a-regression-model/. Accessed on 27 Sept 2018

Press G (2016) Cleaning big data: "most time-consuming, least enjoyable data science task, survey says". Forbes, March 23. Taken from available at: https://www.forbes.com/sites/gilpress/2016/03/23/data-preparation-most-time-consuming-least-enjoyable-data-science-task-survey-says/#5ec6ed56f637. Accessed on 27 Sept 2018

Trochim W, Donnelly J (2006) The research methods knowledge base, 3rd edn. Atomic Dog, Cincinnati

West SG, Finch JF, Curran PJ (1995) Structural equation models with nonnormal variables: problems and remedies. In: Hoyle RH (ed) Structural equation modeling: concepts, issues and applications. Sage, Newbery Park, pp 56–75

Chapter 4
Predictive Models Using Regression

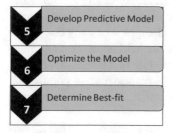

Chapter Learning Objectives

1. Compare and contrast the different types of regression methods.
2. Explain the classical assumptions required in linear regression.
3. Explain how to validate these assumptions and discuss what to do if the assumption is violated.
4. Identify and discuss the components of the multiple linear regression equation.
5. Distinguish the various metrics used to determine the strength of the regression line.
6. Compare and contrast three common variable selection methods.
7. Create regression models using SAS Enterprise Miner™.
8. Evaluate the regression output.

After performing descriptive analysis and data preparation, the next step is to build the predictive model. Regression models can be used as a predictive model. Popular regression models include linear regression, logistic regression, principal component regression, and partial least squares. This chapter defines these techniques and when it is appropriate to use the various regression models. Regression assumptions for each type are discussed. Evaluation metrics to determine model fit including R^2, adjusted-R^2, and p-values are examined. Variable selection techniques (forward, backward, and stepwise) and examination of model coefficients are also discussed. The chapter also provides instruction on implementing regression analysis using SAS Enterprise Miner™ with a focus on evaluation of the output results.

R. V. McCarthy et al., *Applying Predictive Analytics*,
https://doi.org/10.1007/978-3-030-83070-0_4

4.1 Regression

Regression analysis techniques are one of the most common, fundamental statistical techniques used in predictive analytics. The goal of regression analysis is to select a random sample from a population and use the random sample to predict other properties of the population or future events. Regression analysis examines the degree of relationship that exists between a set of input (independent, predictor) variables and a target (dependent) variable. Regression analysis aids organizations in understanding how the target variable (what is being predicted) changes when any one of the input variables change, while the other input variables are held constant. For example, an organization may want to predict their sales revenue in a new market in the southeast (target variable) next year. The organization hypothesizes that population, income level, and weather patterns (input/independent/predictor variables) may influence sales revenue in the new southeast market.

Regression analysis is also known as line-fitting or curve-fitting since the regression equation can be used in fitting a line to data points or in fitting a curve to data points. The regression technique is approached so that the differences in the distance between data points from the line or curve are minimized, referred to as the line of best fit. It is important to note that the relationships between the target variable and the input variable are associative only and any cause-effect is merely subjective.

Regression models involve the following parameters (input variables (X)), target (dependent) variable (Y), and unknown parameters (β). There are numerous existing forms of regression techniques used to make predictions. Several common techniques are linear regression, logistic regression, ordinary least squares (OLS), and partial least squares (PLS). Determination on which model to use depends on three factors—the number of independent variables, the dependent variable, and the shape of the regression line (Fig. 4.1).

4.1.1 Classical Assumptions

Regression analysis is parametric, that is, it comes from a population that follows a probability distribution based on a fixed set of parameters and therefore makes certain assumptions. If these assumptions are violated, then the regression model results will not be suitable for adequate prediction. Therefore, these assumptions must be validated to produce an effective model.

Below are five common assumptions that must be validated for a model to generate good results:

1. *Linearity*: There should be a linear and additive relationship between the target (dependent) variable and the input (predictor) variables.

Fig. 4.1 Determination of regression techniques

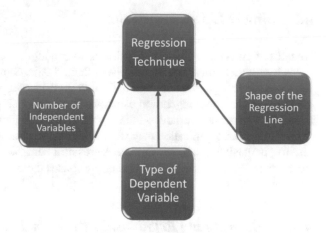

2. *Independence: Multicollinearity* does not exist. Multicollinearity is when two or more input variables are moderately or highly correlated. In linear regression the input variables should not be correlated.
3. *Constant variance* or *homoskedasticity*: An error term (another variable) exists when the predictive model does not precisely portray the relationship between the input and target variables. The error term must have the same variance across all values of the input variables. If the variance is not constant, it is referred to as *heteroskedasticity.*
4. *Autocorrelation*: There should be no correlation between the residual (error) terms. This can occur where the one data point in a variable is dependent on another data point within the same variable. This is frequently seen in time series models. When the error terms are correlated, the estimated standard errors tend to underestimate the standard error.
5. *Normality*: The error terms must be normally distributed for any given value of the input variables with a mean of zero.

4.2 Ordinary Least Squares

Linear regression aims to fit a straight line (or hyperplane for multiple regression) through a data set. *Ordinary least squares or least squares* is a linear regression method that seeks to find out the slope and intercept of a straight line between input variables. It is called least squares as it aims to find out the slope and intercept in such a way as to minimize the sum of the squares of the differences between actual and estimated values of the input variables. Ordinary least squares create a line of best fit by choosing the line where the sum of the distances from the data points to the proposed best-fit line is as small as possible. It is often used synonymously with simple and multiple linear regression; however in advanced statistics, mathematical differences can exist.

4.3 Simple Linear Regression

Recall the purpose of linear regression is to model the relationship between variables by fitting a linear equation to the collected data. Simple linear regression and multiple linear regression are two basic and well-known predictive modeling techniques. Simple linear regression estimates the relationships between two numeric (continuous/interval) variables—the target variable (Y) and the input variable (X). For example, an organization may want to relate the number of clicks on their website to purchasing a product. Can the regression equation predict how many clicks on average it takes before the customer purchases the product?

4.3.1 Determining Relationship Between Two Variables

It is imperative to ensure that a relationship between the variables is determined prior to fitting a linear model to the collected data. However, as mentioned earlier, it is important to note that just because a relationship exists does not mean that one variable causes the other; rather some significant association between the two variables exists. A common method to determine the strength of the relationship between two variables is a *scatter plot*. If no association between the target and input variables exists, then fitting a linear regression model to the collected data will not provide a suitable model. A numerical measure of association between two variables is the *(Pearson) correlation coefficient (r)*. Correlation coefficients range between -1 and $+1$. A -1 one correlation coefficient suggests the data points on the scatter plot lie exactly on a descending straight line, while a $+1$ correlation coefficient suggests the data points on the scatter plot lie exactly on an ascending straight line. A correlation of 0 indicates no linear relationship exists. (See Chap. 3, Fig. 3.22, for examples of scatter plots with negative, positive, and no correlation.) The closer the correlation coefficient (r) is to either $+1$ or -1, the greater the association strength. Keep in mind that correlations are very sensitive to outliers.

4.3.2 Line of Best Fit and Simple Linear Regression Equation

The goal of simple linear regression is to develop an equation that provides the *line of best fit*. The line of best fit is a straight-line equation that minimizes the distance between the predicted value and the actual value. Figure 4.2 shows a scatter plot with a line of best fit also known as the *regression line* or *trend line*. Software programs such as SAS Enterprise Miner™ compute the line of best fit using a mathematical process known as *least squares estimation,* that is, the program will calculate the line of best fit such that the squared deviations of the collected data points are minimized (i.e., ordinary least squares).

Fig. 4.2   Scatter plot with line of best fit

The simple linear regression equation is

$$\text{Outcome}_t = \text{model}_t + \text{error}$$
$$Y_t = \beta_0 + \beta_1 X_{1t} + \varepsilon$$

Whereby
Y_t—is the target variable (what is being predicted).
X_{1t}—is the input variable.
β_0—is the intercept of the regression line. Also referred to as the constant.
β_1—is the slope of the regression line (beta coefficient).
ε—is the error.

Figure 4.3 provides a clear visual of the error term. The dots are the actual plotted values from the collected data. The error term is the difference between the actual plotted values and the predicted value from the regression equation.

4.4   Multiple Linear Regression

Multiple linear regression analysis is used to predict trends and future values. Like simple linear regression, multiple linear regression estimates the relationship between variables. The key difference is that multiple linear regression has two or more input variables. Therefore, multiple linear regression estimates the relationship between a numeric (continuous/interval) target variable and two or more input variables. The input variables can be either numeric or categorical.

The multiple linear regression equation is

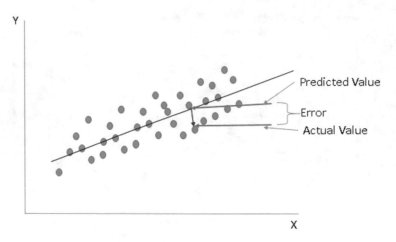

Fig. 4.3 Illustration of the error term

$$Y_t = \beta_0 + \beta_1 X_{1t} + \beta_2 X_{2t} + \cdots + \beta_k X_{kt} + \varepsilon$$

Whereby,

Y_t—is the target variable (what is being predicted).

$X_{1t}, X_{2t}, \ldots, X_{kt}$—are the input variables explaining the variance in Y_t.

β_0—is the intercept of the regression line. Also referred to as the constant.

$\beta_1 + \beta_2 + \cdots + \beta_k$—are the slopes for each input corresponding input variable and the regression coefficients.

ε—is the error.

4.4.1   Metrics to Evaluate the Strength of the Regression Line

R^2 (r-squared) or coefficient of determination is a commonly used metric of the strength of the relationship between the variables. R^2 provides the percentage variation in the target (Y) variable explained by the input (Y) variables. The range is from 0% to 100%. Recall that the correlation coefficient tells how strong of a linear relationship there is between two variables. R^2 is simply the square of the correlation coefficient. Typically, the higher the R^2, the more robust the model is.

Adjusted R^2 or the adjusted coefficient of determination: As more variables are added to the model, the R^2 increases. However, not all variables added to the model are significant and fit the model. Therefore, the R^2 can be misleading. The adjusted R^2 overcomes this weakness by making an adjustment to the R^2 by integrating the model's degrees of freedom. When an input variable has a correlation to the predictor variable, then an increase is made to the adjusted R^2. Conversely, when an input variable does not have a strong correlation with the predictor variable, then a decrease is made to the adjusted R^2. Hence, the adjusted R^2 penalizes the R^2 for including variables that do not improve the fit of the model. The adjusted R^2 will always be lower than the R^2 and could in fact be negative. When using multiple regression, always use the adjusted R^2 rather than R^2.

P-value: The *p*-value (probability value) is another common metric used to determine the significance of the model results when applying hypothesis testing. The *p*-value is a number between 0 and 1. Generally, a *p*-value ≤ 0.05 (considered a small *p*-value) indicates strong evidence against the null hypothesis (i.e., the assumption that is attempting to be tested); therefore the null hypothesis would be rejected. A *p*-value >0.05 (considered a large *p*-value) indicates weak evidence against the null hypothesis; therefore, the null hypothesis would fail to be rejected. A *p*-value close to the 0.05 cutoff would be considered marginal and could go either way. In testing whether a linear relation exists, the following would be the null and alternative hypothesis:

H_0: No linear relationship exists.
H_A: Linear relationship does exist.

If the *p*-value is small (≤ 0.05), then the result is considered statistically significant and something happened. In other words, it is greater than chance alone. So the null (H_o) hypothesis would be rejected, and the alternative (H_A) hypothesis would be supported. That is, a linear relationship does exist. The *p*-value may also be referred to as alpha (α).

4.4.2   Best-Fit Model

Like other linear regression models, multiple linear regression looks to identify the best-fit model. As previously mentioned, adding input variables to a multiple linear regression model, the amount of explained variance in the dependent variables (R^2) always increases. However, adding too many input variables without justification can result in an overfit model with misleading R^2s, regression coefficients, and *p*-values. That is, an overfit model results when the model is too complex.

4.4.3   Selection of Variables in Regression

Carefully selecting the input variables can improve model accuracy, avoid overfitted models, reduce complexity of the model, generate easier interpretation of the model, and train the model faster. In Chap. 3 variable clustering and principal component analysis were discussed as two methods to reduce variables. There are several sequential methods to use in the variable selection with the common methods being forward selection, backward selection, and stepwise selection. Each is discussed below.

Forward selection is a simple method for variable selection. In this approach variables are added to the model one at a time based upon a pre-set significance level (*p*-value or "alpha to enter"). The process begins by adding the most significant variable to the model. The process then continues to add variables that are significant until none of the remaining variables are significant. The pre-set *p*-value is typically greater than the standard 0.05 level. Many software packages are less restrictive and have default values set at 0.10 or 0.15. The forward selection method may be useful when multicollinearity may be a problem. However, a limitation of the forward selection process is that each addition of a new variable may cause one or more of the included variables to be non-significant.

Backward selection is the opposite of forward selection and may help overcome the limitation noted above. The backward selection process starts with fitting the model with all identified input variables. In this approach variables are removed from the model one at a time based upon a pre-set significance level (*p*-value or "alpha to remove"). The process begins by removing the least significant variable to the model. The process then continues to remove variables that are not significant until all remaining variables in the model are significant. A limitation of this method is that the model selected may include unnecessary variables.

Stepwise selection incorporates both forward and backward selection processes. The basic idea behind the stepwise selection method is that the model is developed from input variables that are added or removed in a sequential manner into the model until there is no arguable reason to add or remove any more variables. The stepwise selection method starts with no input variables in the model. After each step where a variable is added, a step is performed to see if any of the variables added to the model have had their significance reduced below the pre-set significance level. Two significant levels are determined. A significance level ("alpha to enter") is set for identifying when to add a variable into the model, and a significance level ("alpha to remove") is set for identifying when to remove a variable from the model. As mentioned earlier, the significance levels are typically greater than the standard 0.05 level as it can be too restrictive. Software packages will commonly have a default of 0.05 to 0.15 significance levels.

These three sequential methods should produce the same models if there is very little correlation among the input variables and no outlier problems. A serious limitation with these methods exists with small data sets. Small data sets can lead to inadequate prediction models with large standard errors. The models are inadequate as there is insufficient data to identify significant associations between the target and input variables. Several "common rules of thumb" have concluded that anywhere between 5 to 20 data points per input variable is required. More importantly though, it is not the number of data points that determines a useful regression model; rather it is whether the model satisfies the five common assumptions—linearity, independence, constant variance, autocorrelation, and normality.

4.5   Principal Component Regression

4.5.1   Principal Component Analysis Revisited

When there are many input variables, it may be necessary to reduce the number of input variables to permit more meaningfully interpretation of the data. In Chap. 3, principal component analysis (PCA), a non-parametric analysis, was described as another variable reduction strategy and used when there are many redundant variables or variables that are possibly correlated. PCA is frequently used in exploratory data analysis but can also be used to build predictive models.

PCA applies a method referred to as *feature extraction*. Feature extraction creates "new" input variables. The "new" input variables are a combination of each of the "old" input variables. The new input variables are created through a process known as *orthogonal transformation* (a linear transformation process). The new input variables (referred to as principal components) are now linearly uncorrelated variables. Orthogonal transformation orders the new input variables such that the first principal component explains the largest amount of variability in the data set as possible, and each subsequent principal component in sequence has the highest variance possible under the limitation that it is orthogonal (perpendicular) to the preceding components. Since the new input variables are ordered from the highest variance to the smallest variance (least important), a decision can be made which variables to exclude from the prediction model. The results of PCA are normally defined as component scores or factor scores.

4.5.2   Principal Component Regression

Principal component regression is an extension of PCA. It takes the untransformed target variable and regresses it on the subset of the transformed input variables (principal components) that were not removed. This is possible since the newly transformed input variables are independent (uncorrelated) variables.

4.6   Partial Least Squares

Multiple linear regression is a good regression technique to use when data sets contain easy-to-deal with input variables to explain the predictor variables, the input variables are not significantly redundant (collinear), the input variables have an implied relationship to the predictor variable, and there are not too many input variables (e.g., the number of observations does not exceed the number of input variables). Conversely, if these conditions fail, multiple linear regression would be inappropriate and would result in a useless, overfit predictive model.

Whereas *partial least squares (PLS)* is a flexible regression technique that has features of principal component analysis and extends multiple linear regression. PLS is best used when the data set contains fewer observations than input variables and high collinearity exists. PLS can also be used as an exploratory analysis tool to identify input variables to include in the model and outliers. Like multiple linear regression, PLS' overarching goal is to create a linear predictive model.

4.7 Logistic Regression

The regression models discussed so far have a numeric (continuous) target variable, and the data set is normally distributed. Generalized linear models are an extension of traditional linear models and are used when the target variable is discrete and the error terms do not follow a normal distribution. Logistic regression belongs in this category of regression models and is one of the most commonly used techniques for discrete data analysis.

In predictive analytics, logistic regression is used when examining the relationship between a categorical (class) target variable and one or more input variables. The target variable can be either binary, multi-nominal, or ordinal. The input variables can be either continuous (numeric) or categorical. Logistic regression aids in estimating the probability of falling into a certain level of the categorical (class) target output given a set of input variables. Simply stated, logistic regression provides the likelihood or probability a particular outcome/event will occur. Logistic regression predicts the value of the outcome. For example:

- How likely will a customer buy another Apple product if they already own an iPhone?
- How likely will a frequent flyer continue to be loyal to the airline if the airline charges for carry-on luggage?
- What is the probability to get hired at a Big Four accounting firm with a GPA greater than 3.5 at a particular university?
- What is the likelihood a homeowner will default on a mortgage?
- What is the probability a customer will cancel a subscription?

Logistic regression violates the following linear regression assumptions:

1. Residuals are not normally distributed.
2. Constant variance of the error terms.

Some assumptions exist in logistic regression. The assumptions are:

1. Adequate sample size.
2. Multicollinearity does not exist.
3. Errors need to be independent but do not have to be normally distributed.
4. There are no outliers.

The nature of the categorical (class) target variable determines which of the three types of logistic regression will be used—binary logistic regression, multi-nominal logistic regression, and ordinal logistic regression.

4.7.1 Binary Logistic Regression

Binary logistic regression is used when the target variable is binary. In other words, it has two possible outcomes. Examples include passing or failing a course, answering yes or no to a question, win or lose a game, and an insurance claim is fraudulent or not fraudulent. Binary logistic regression can be either univariant (one categorical target variable and one input variable) or multivariant (one categorical target variable and two or more input variables). Binary logistic regression predicts the odds of an occurrence based on the values of the input variables.

As mentioned, logistic regression violates the assumptions of linear regression particularly the assumption that the residuals are normally distributed. The reason for this is that logistic regression is used for predicting target variables that belong in one of a finite set of categories rather than a target variable that has a continuous outcome. To overcome this violation binomial logistic regression calculates the odds of the occurrence happening at different levels of each of the input variables and then takes its logarithm to create a continuous condition as a transformed type of the target variable. Because the target variable is finite (e.g., 1 if the condition exists; 0 if the condition does not exist), graphing a binary logistic regression with one input variable results in an s-shaped (sigmoid curve) graph (Fig. 4.4).

To convert a binary variable to a continuous variable, logistic regression starts with an equation that predicts the probability that an event will happen and then calculates the odds the event will happen (Eq. 4.1), and finally the equation takes the

Fig. 4.4 Binary logistic regression distribution (univariant)

$$P(Y) = \frac{\exp(\beta_0 + \beta_1 X_1)}{1 + \exp(\beta_0 + \beta_1 X_1)}$$

Eq. 4.1 Univariate binary logistic equation

$$P(Y) = \frac{\exp(\beta_0 + \beta_1 X_1 + \cdots + \beta_i X_i)}{1 + \exp(\beta_0 + \beta_1 X_1 + \cdots + \beta_i X_i)}$$

$$= \frac{\exp(\mathbf{X}\beta)}{1 + \exp(\mathbf{X}\beta)}$$

$$= \frac{1}{1 + \exp(-\mathbf{X}\beta)}$$

Eq. 4.2 Multivariant binary logistic equation

logarithm of the odds to transform the target variable to a continuous criterion, thus a transformed target variable (Eq. 4.2). The binary logistic regression equation for one input variable (univariant) and one target is.

The equation for a multivariant binary logistic regression is.

Where $P(Y)$ is the probability of the binary predictor variable (Y) successfully occurring ($Y = 1$). Model estimates of the probability will always be between 0 and 1. In ordinary least squares regression, changing an input (X) variable is the same everywhere. However, in logistic regression, changing an input (X) variable depends on the values of the other input (X) variables; therefore changing an input (X) variable is not the same. This is due to the relationship between the input (X) variables, and the probability of success is nonlinear.

B_0 is the constant and β_1 through β_p are the coefficients.

Essential to logistic regression is the ratio of the outcome probabilities or the odds ratio:

$$\text{Odds ratio} = \frac{P(Y)}{1 - P(Y)}$$

The odds ratio is not the same as the probability an event will occur. An easy example to understand this is that if an event occurs in 2 of 10 events (20% chance), the odds ratio is 0.2/(1–0.2) which equals 0.2/0.8 = 0.25, and the odds of the event not occurring is 0.8/0.2 = 4.0. Stated as the odds of the event not occurring is four times greater than the event occurs.

An odds ratio of 1 indicates there is no association between the target and input variable.

There are other equivalent algebraic equations or scales to write the logistic model. The previous equation provides the probability of the event. A second equation provides the odds in favor of the event occurring.

$$\frac{P(Y)}{1-P(Y)} = \exp\left(\beta_0 + \beta_1 X_1 + \cdots + \beta_i X_i\right)$$

The outcome of the odds in favor of an event occurring is a number between 0 and ∞. Odds of 0 concludes with certainty the event does not occur. Odds of 1 concludes a probability of 1:2 chance, and an odds of ∞ concludes with certainty the event occurs.

The third equation (or scale) is the log-odds equation. The equation expresses the (natural) logarithm of the odds is a linear function of the input (X) variables. It is also referred to as the logit transformation of the probability of success:

$$\log\left(\frac{P(Y)}{1-P(Y)}\right) = \exp\left(\beta_0 + \beta_1 X_1 + \cdots + \beta_i X_i\right)$$

The log-odds (natural logarithm of the log-odds or logit) equation is the key equation for logistic regression. The log-odds equation results in a number between $-\infty$ and ∞. A log-odds of $-\infty$ concludes with certainty the event does not occur. A log-odds of 0 concludes the probability is 1:2, and a log-odds of ∞ concludes with certainty the event occurs. Therefore, an increase in the log-odds relates to an increased certainty that the event will occur.

4.7.2 Examination of Coefficients

To assess the contribution of the individual input variables, the regression coefficients need to be examined. Two tests used to evaluate the significance of the input variables are the Wald test and the odds ratio.

Wald test: The Wald test is a measure used to evaluate if the model's input variables are significant. It is calculated by taking the square of the regression coefficient divided by the square of the standard error of the coefficient. If the variable is significant (i.e., the Wald test results are not equal to zero), then it adds to the model. If the variable is not significant (i.e., the Wald test equals zero), then the variable can be deleted from the model without harming the fit of the model.

Odds ratio: In linear regression, the regression coefficients denote the change in the input variable for each unit change in the target variable. However, this is not the case with logistic regression. In logistic regression, the regression coefficients denote the change in the logit for each unit change in the input variable. For the odds ratio, the exponential function of the regression coefficient is used. Many predictive modeling software packages provide an output table with the odds ratio for the input variables. The values in the odds ratio indicate the percentage change in the odds change for a one-unit change in the input variable. When the odds ratio is greater than 1, a positive relationship exists. Conversely, if the odds ratio is less than 1, a negative relationship exists.

Model Validation (Model Fit):
There are several tests that can be used to validate the logostic regression model's overall goodness of fit. Each produces a p-value. These are tests of the null hypothesis that the fitted model is correct. If the p-value is low (e.g., <0.05), then the model is rejected. If the p-value is high (e.g., >0.05), then the model is accepted. These tests include:

- Deviance test
- Pearson Chi-square
- Hosmer-Lemeshow test

4.7.3   Multinomial Logistic Regression

Multinomial logistic regression is used when the target variable has three or more categories (classes) with no regular ordering to the levels. Examples include color (green, blue, red) and stores (Walmart, Target, JC Penney), which search engine (Yahoo!, Google, Bing). The multinomial model would predict the chance of selecting a particular category (class).

Assumptions used in multinomial logistic regression are:

1. Each input variable has a single value for each observation.
2. The predictor variable cannot be exactly predicted from the input variables for each observation.
3. While the input variables do not need to be statistically independent, it is assumed that collinearity is relatively low.

4.7.4   Ordinal Logistic Regression

Ordinal logistic regression is used when there are three or more categories (classes) with a normal ordering to the data. The ranking of the levels does not mean the intervals between the levels are equal. Examples could include the Likert ratings on a survey scale of 1–5 or an employment status such as full time, part time, or unemployed.

4.8   Implementation of Regression in SAS Enterprise Miner™

Let's examine the use of regression analysis as a predictive analytics tool. There are four nodes in SAS Enterprise Miner™ that model regression including:

1. Regression: This includes both linear and logistic regression.
2. Dmine regression: This regression model computes a forward stepwise least squares regression.
3. Partial least squares (PLS).
4. LARS (least angle regressions).

The first example will use the automobile insurance claim data set to demonstrate a logistic regression analysis (Fig. 4.5). Linear and logistic regressions are performed through the use of the regression node. SAS Enterprise Miner™ will select linear regression as a default when the target variable is an interval and logistic regression as a default when the target variable is nominal or ordinal.

Now let's examine the key properties in the regression node (Fig. 4.6) for creating a linear or logistic regression.

4.8.1   Regression Node Train Properties: Class Targets

Regression type specifies the type of regression you want to run. Logistic regression is the default for ordinal and binary targets. Linear regression is the default for interval targets. The logistic regression attempts to predict a probability that the target will acquire the event of interest. The linear regression computes a value of a continuous target:

- *Link function*—For linear regression, the link function $g(M) = X\beta$.

- For logistic regression, you can select:

    - *Cloglog* (complementary log-log)—This link function is the inverse of the cumulative extreme-value function.
    - *Logit*—This link function is the inverse of the cumulative logistic distribution function (and is the default).
    - *Probit*—This link function is the inverse of the cumulative standard normal distribution.

Fig. 4.5 Regression node

Fig. 4.6 Regression node
properties

.. Property	Value
Node ID	Reg
Imported Data	
Exported Data	
Notes	
Train	
Variables	
□Equation	
⁝Main Effects	Yes
⁝Two-Factor Interactions	No
⁝Polynomial Terms	No
⁝Polynomial Degree	2
⁝User Terms	No
⁝Term Editor	
□Class Targets	
⁝Regression Type	Logistic Regression
⁝Link Function	Logit
□Model Options	
⁝Suppress Intercept	No
⁝Input Coding	Deviation
□Model Selection	
⁝Selection Model	Stepwise
⁝Selection Criterion	Validation Misclassification
⁝Use Selection Defaults	Yes
⁝Selection Options	

4.8.2   Regression Node Train Properties: Model Options

- *Suppress intercept*—When set to Yes, this will suppress the intercepts when coding class variables.
- *Input coding*—Used to specify the method for coding class variables.

  - *GLM*—Generalized linear models (GLM) coding, also known as dummy coding, uses a parameter estimate to estimate the difference between each level and a reference level.
  - *Deviation*—Uses parameters to estimate the difference between each level and the average across each level. This is the default.

4.8.3   Regression Node Train Properties: Model Selection

- *Selection model*—Specifies whether backward, forward, or stepwise selection will be utilized. In addition, none (which is the default) could be specified. When none is chosen, all inputs are fit to the regression model.

- *Selection criterion*—When forward, backward, or stepwise selection method is chosen, this property specifies the criterion for choosing the final model. The options include:

  - *Default*
  - *None*—Chooses standard variable selection based on the *p*-values.
  - *Akaike's Information Criterion* (AIC)—Chooses the model with the smallest criterion.
  - *Schwarz's Bayesian Criterion* (SBC)—Chooses the model with the smallest criterion.
  - *Validation Error*—Uses the smallest error rate from the validation data set. For logistic regression the error rate is the negative log-likelihood. For linear regression it is the sum of square errors.
  - *Validation Misclassification*—Uses the smallest misclassification rate for the validation data set.
  - *Profit/Loss*—If a profit/loss matrix is defined, this uses the training data profit/loss.
  - *Validation Profit/Loss*—If a profit/loss matrix is defined, this uses the validation data profit/loss.

Using the automobile insurance claim fraud data set, let's examine the results of a stepwise logistic regression. Logistic regression is used because the target variable is binary. The link function will use the default (logit), and the selection criteria will be validation misclassification. Upon completion of the model run, examine the fit statistics window (Fig. 4.7). The model shows an average squared error of 0.05813 on the Validation partition. This will be used to compare the stepwise selection regression model to other models. Note that the difference in the average squared error between the Train and Validation partitions is not significant. When significant differences exist, it indicates a possible overfit of the regression model.

The first step (Step 0) in a stepwise regression model is the intercept only. Subsequent steps then add in the input variables. In this example (Fig. 4.8), the variable Gender is added to the regression model. The hypothesis that the two models (Step 0 and Step1) are identical is tested. Note that the Pr > ChiSq is <0.0001. A value that is close to 0 indicates that it is a significant input. A value that is close to 1 indicates it is an extraneous input. Gender, in this example, is a significant input. Inputs that are highly significant should be included in further analysis, and inputs that are extraneous (i.e., insignificant) can be removed. The odds ratio estimate indicates the factor by which the input variable increases or decreases the primary outcome odds estimate of the target variable. It is obtained by exponentiation of the parameter of the input.

Fit Statistics						
Target	Target Label	Fit Statistics	Statistics Label	Train		Validation
Fraudulent_Claim	Fraudulent_Claim	_AIC_	Akaike's Information Criterion	1379.884		
Fraudulent_Claim	Fraudulent_Claim	_ASE_	Average Squared Error	0.057333		0.05813
Fraudulent_Claim	Fraudulent_Claim	_AVERR_	Average Error Function	0.229877		0.232358
Fraudulent_Claim	Fraudulent_Claim	_DFE_	Degrees of Freedom for Error	2996		
Fraudulent_Claim	Fraudulent_Claim	_DFM_	Model Degrees of Freedom	1		
Fraudulent_Claim	Fraudulent_Claim	_DFT_	Total Degrees of Freedom	2997		
Fraudulent_Claim	Fraudulent_Claim	_DIV_	Divisor for ASE	5994		4002
Fraudulent_Claim	Fraudulent_Claim	_ERR_	Error Function	1377.884		929.8986
Fraudulent_Claim	Fraudulent_Claim	_FPE_	Final Prediction Error	0.057371		
Fraudulent_Claim	Fraudulent_Claim	_MAX_	Maximum Absolute Error	0.938939		0.938939
Fraudulent_Claim	Fraudulent_Claim	_MSE_	Mean Square Error	0.057352		0.05813
Fraudulent_Claim	Fraudulent_Claim	_NOBS_	Sum of Frequencies	2997		2001
Fraudulent_Claim	Fraudulent_Claim	_NW_	Number of Estimate Weights	1		
Fraudulent_Claim	Fraudulent_Claim	_RASE_	Root Average Sum of Squares	0.239442		0.241101
Fraudulent_Claim	Fraudulent_Claim	_RFPE_	Root Final Prediction Error	0.239522		
Fraudulent_Claim	Fraudulent_Claim	_RMSE_	Root Mean Squared Error	0.239482		0.241101
Fraudulent_Claim	Fraudulent_Claim	_SBC_	Schwarz's Bayesian Criterion	1385.889		
Fraudulent_Claim	Fraudulent_Claim	_SSE_	Sum of Squared Errors	343.6517		232.635
Fraudulent_Claim	Fraudulent_Claim	_SUMW_	Sum of Case Weights Times Freq	5994		4002
Fraudulent_Claim	Fraudulent_Claim	_MISC_	Misclassification Rate	0.061061		0.061969

Fig. 4.7 Fit statistics

```
        Likelihood Ratio Test for Global Null Hypothesis: BETA=0

     -2 Log Likelihood              Likelihood
   Intercept     Intercept &           Ratio
      Only       Covariates         Chi-Square      DF      Pr > ChiSq

    1377.884      1308.115            69.7686         1        <.0001


                Type 3 Analysis of Effects

                                   Wald
   Effect             DF        Chi-Square     Pr > ChiSq

   Gender              1          56.6360        <.0001


                        Analysis of Maximum Likelihood Estimates

                                        Standard      Wald                    Standardized
   Parameter           DF    Estimate     Error     Chi-Square   Pr > ChiSq     Estimate     Exp(Est)

   Intercept            1     -2.9517     0.0939      987.33       <.0001                      0.052
   Gender         F     1     -0.7070     0.0939       56.64       <.0001                      0.493


                     Odds Ratio Estimates

                                              Point
   Effect                                   Estimate

   Gender        F vs M                       0.243
```

Fig. 4.8 Automobile insurance claim fraud example regression output

4.9 Implementation of Two-Factor Interaction and Polynomial Terms

In addition to executing either a linear or logistic regression function with the input variables, it is possible to set up two-factor interactions or polynomial terms to investigate the interaction of two or more input variables simultaneously against a target variable. To do so there are several properties that should be addressed (refer to Fig. 4.5). These include:

4.9.1 Regression Node Train Properties: Equation

- *Two-Factor Interactions*—When set to Yes, the Two-Factor Interactions property will include all two-factor interactions for class variables that have a status of Use.
- *Polynomial Terms*—When set to Yes, the Polynomial Terms property will include polynomial terms for interval variables with status of Use. An integer value for the Polynomial Degree property must be specified.
- *Polynomial Degree*—When the Polynomial Terms property is set to Yes, this property specifies the highest degree of polynomial terms for interval variables with status set to Use. The Polynomial Degree property can be set to 1 or 2.
- *User Terms*—When this is set to Yes, it will create a Terms data set using the Term Editor.
- *Term Editor*—The Term Editor is used to create two factor interactions for class variables and polynomial terms for interval variables for variables with a status of Use.

The automobile insurance claim fraud example can be extended to include a two-factor interaction term. In this case, let's examine the interaction between gender and employment status (Fig. 4.9).

Interaction terms result in a detailed analysis of the categories within each variable and their impact on the target variable (Fig. 4.10). Note that the interaction between female (gender) and employed is significant, while the interaction between female and retired is not.

Fig. 4.9 Two-factor interaction

Type 3 Analysis of Effects

Effect	DF	Wald Chi-Square	Pr > ChiSq
Employment_Status*Gender	4	76.0170	<.0001

Analysis of Maximum Likelihood Estimates

Parameter		DF	Estimate	Standard Error	Wald Chi-Square	Pr > ChiSq	Standardized Estimate	Exp(Est)
Intercept		1	-3.0343	0.1000	920.55	<.0001		0.048
Employment_Status*Gender	Disabled F	1	0.2625	0.3381	0.60	0.4375		1.300
Employment_Status*Gender	Employed F	1	-1.0261	0.1183	75.24	<.0001		0.358
Employment_Status*Gender	Medical Leave F	1	0.7221	0.3333	4.70	0.0302		2.059
Employment_Status*Gender	Retired F	1	0.0771	0.3585	0.05	0.8298		1.080

Fig. 4.10 Two-factor interaction term sample output

Fig. 4.11 DMINE regression node

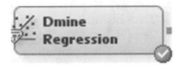

4.10   DMINE Regression in SAS Enterprise Miner™

A DMINE node (Fig. 4.11) provides a quick method to compute a forward stepwise least squares regression for the purpose of identifying the input variables that are useful predictors of a target variable. The logistic function uses only one regression variable and calculates two parameter estimates: the intercept and the slope. For categorical input variables, missing values are considered to be a new category. For interval input variables, missing values are replaced with the mean. If there are any missing values in the target variable, that observation is excluded from analysis.

The DMINE procedure selects an input variable that contributes maximally to the R-squared value in each step that it computes. Optionally the two-factor interactions can also be tested. Another option permits the testing of nonlinear relationship between interval input variables and the target variable by binning the interval inputs.

4.10.1   DMINE Properties

Let's examine the key properties of the DMINE regression node (Fig. 4.12).

DMINE Regression Node Train Properties
- *Maximum Variable Number*—Sets the upper limit for the number of input variables to be considered in the regression model. The default setting is 3000.

DMINE Regression Node Train Properties: R-Squared Options
- *Minimum R-Squared*—Sets the lower limit for the R-squared value of an input variable to be included in the regression model.
- *Stop R-squared*—Sets the lower limit for the lowest cumulative R-squared value.

Fig. 4.12 DMINE
properties

.. Property	Value
General	
Node ID	DmineReg
Imported Data	
Exported Data	
Notes	
Train	
Variables	
Maximum Variable Number	3000
R-Square Options	
Minimum R-Square	0.005
Stop R-Square	5.0E-4
Created Variables	
Use AOV16 Variables	Yes
Use Group Variables	Yes
Use Interactions	No
Print Option	Default
Use SPD Engine Library	Yes
Status	
Create Time	5/11/18 10:45 AM
Run ID	6005f259-d095-464e-8de0
Last Error	
Last Status	Complete
Last Run Time	9/19/18 5:02 PM
Run Duration	0 Hr. 0 Min. 17.29 Sec.
Grid Host	
User-Added Node	No

DMINE Regression Node Train Properties: Created Variables

- *Use AOV16 Variables*—When set to Yes, interval variables are binned into 16 equally spaced groups. The purpose of binning is to identify nonlinear relationships between the interval inputs and the target variable.
- *Use Group Variables*—When set to Yes, this will attempt to reduce the number of levels for class variables.
- *Use Interactions*—When set to Yes, two-factor interactions of input variables will be evaluated in the regression model.
- *Print Option*—Controls how much detail is included in the output.

  - *Default*—Suppresses some of the detail.
  - *All*—Includes all of the detail.
  - *None*—Suppresses all output.

- *Use SPDE Library*—Utilizes the SAS Scalable Performance Data Engine for larger volume analysis.

Target	Target Label	Fit Statistics	Statistics Label	Train	Validation
Fraudulent_Claim	Fraudulent_Claim	_ASE_	Average Squared Error	0.054139	0.056494
Fraudulent_Claim	Fraudulent_Claim	_DIV_	Divisor for ASE	5994	4002
Fraudulent_Claim	Fraudulent_Claim	_MAX_	Maximum Absolute Error	0.982669	0.982669
Fraudulent_Claim	Fraudulent_Claim	_NOBS_	Sum of Frequencies	2997	2001
Fraudulent_Claim	Fraudulent_Claim	_RASE_	Root Average Squared Error	0.232677	0.237685
Fraudulent_Claim	Fraudulent_Claim	_SSE_	Sum of Squared Errors	324.508	226.089
Fraudulent_Claim	Fraudulent_Claim	_DISF_	Frequency of Classified Cases	2997	2001
Fraudulent_Claim	Fraudulent_Claim	_MISC_	Misclassification Rate	0.061061	0.061969
Fraudulent_Claim	Fraudulent_Claim	_WRONG_	Number of Wrong Classifications	183	124

Fig. 4.13 DMINE fit statistics

```
The DMINE Procedure


                  Effects Chosen for Target: Fraudulent_Claim


                                                                           Sum of
Effect                              DF      R-Square    F Value  p-Value    Squares

Group: Vehicle_Class                 2      0.022068   33.780883  <.0001    3.791809
Class: Gender                        1      0.021111   66.036484  <.0001    3.627409
AOV16: Claim_Amount                 13      0.004091    0.984365   0.4639   0.702978
```

Fig. 4.14 DMINE output results

4.10.2 DMINE Results

Using the automobile insurance claim fraud data set, let's examine the results of a DMINE regression analysis (Fig. 4.13). Note that the average squared error is equal to 0.055494. This represents an improvement compared to the stepwise logistic regression average squared error of 0.05813. Also note the difference between the train and validation partitions' average squared errors is relatively small indicating the model does not appear to be overfitting.

The output of the DMINE regression node provides a list of the variables that were used in the regression model (Fig. 4.14). In this example, vehicle class, gender, and claim amount were included in the final regression model. Vehicle class and gender are highly significant (based upon their p-values), and claim amount is moderately significant, accounting for less of the overall contribution to the final regression model.

Cumulative lift provides a visual of the strength of the model compared to random guessing. Random guessing or no lift is equal to 1. The higher the cumulative lift, the better the predictability of the model. Looking at the cumulative lift output results (Fig. 4.15), the depth (x-axis) is the cumulative percentage of data evaluated. The cumulative lift (y-axis) shows how far the model is from random guessing. Generally, when analyzing the cumulative lift graph, it is important to see how the model performs in the early stages of running the model. The higher the lift, the better. For example, at the 20% depth, the cumulative lift is 2.1 indicating the model is two+ times better than random guessing. In this model, the cumulative lift remains at that level for 40% of the data, and therefore we should continue to consider this model until a better model is developed.

Fig. 4.15 DMINE cumulative lift

4.11   Partial Least Squares Regression in SAS Enterprise Miner™

A partial least squares (PLS) regression node (Fig. 4.16) provides an alternative method to multiple linear regression to analyze data sets that have a large number of input variables and a single continuous or binary target.

PLS is also useful when there is significant collinearity between the input variables. This technique tends to reduce the impact of overfit models when a small number of variables represent a significant amount of the prediction of the target variable (i.e., the variation in the response of the target). This is accomplished by reducing the set of input variables to principal component matrices that are then used to predict the value of the target variable (SAS Enterprise Miner Reference Help 2018). PLS excludes observations with missing values.

4.11.1   Partial Least Squares Properties

Let's examine the key properties of the partial least squares regression node (Fig. 4.17).

Partial Least Squares Node Train Properties: Modeling Techniques
- *Regression Model*—Specifies the technique that will be used; options include:

   – *PLS*—(Default) uses the original partial least squares method.

- *PCR*—Uses principal component regression to perform general factor extraction. The *k* components (that have the highest eigenvalues) are used to predict the target variable.

Fig. 4.16 Partial least
squares node

Fig. 4.17 Partial least
squares properties

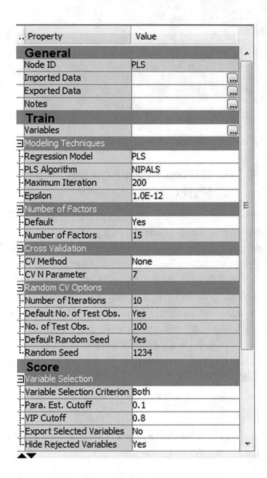

- *SIMPLS*—Used when there are multiple target variables. When you have a
  single target, this is identical to PLS.

- *PLS Algorithm*—Specifies the algorithm used when the regression model is PLS
  or SIMPLS; options include:

  - *NIPALS*—(Default) uses a nonlinear, iterative, partial least squares algorithm.
  - *SVD*—Uses a singular value decomposition algorithm that is the most accu-
    rate but least computationally efficient.
  - *Eigenvalue*—Uses an eigenvalue decomposition algorithm.

- *RLGW*—Uses an iterative approach algorithm that is best used when there are many input variables.

- *Maximum Iteration*—Specifies the maximum number of iterations for the NIPALS or RLGW algorithms. The default is 200.
- *Epsilon*—Specifies the convergence criterion for the NIPALS or RLGW algorithms.

Partial Least Squares Node Train Properties: Number of Factors
- *Default*—When set to No, this enables you to set the number of factors property (below).
- *Number of Factors*—Specifies the total number of factors to be extracted.

Partial Least Squares Node Train Properties: Cross Validation
- *Cross Validation Method*—Specifies the method used for cross validation checks to validate the PLS model; options include:

  - *One at a time*—Uses a single observation at a time.
  - *Split*—Drops every nth observation.
  - *Random*—Drops a random subset of observations.
  - *Block*—Drops groups of n observations.
  - *Test Set*—Uses either the validation data set or the test data set.
  - *None*—No cross validation is performed.

Partial Least Squares Node Train Properties: Variable Selection
- *Variable Selection Criterion*—Specifies which variable selection criterion will be used; options include:

  - *Coefficient*—Selects variables based on parameter estimates.
  - *Variable Importance*—Selects variables based upon variable importance.
  - *Both*—Selects variables using parameter estimates and variable importance.
  - *Either*—Selects variables using parameter estimates or variable importance.

4.11.2   Partial Least Squares Results

Using the automobile insurance claim fraud data set, let's examine the results of the partial least squares regression analysis (Fig. 4.18). Note that the average squared error is equal to 0.055189. This represents a slight improvement compared to the DMINE regression. Also note the difference between the train and validation partitions' average squared errors is still also relatively small indicating the model does not appear to be overfitting.

The partial least squares node output includes an absolute standardized parameter estimate bar graph (Fig. 4.19) which provides a color graphic of the parameter estimates for each variable. Red bars indicate a negative value, and blue bars denote a positive value. In SAS Enterprise Miner™, each bar contains a tooltip text that displays the actual value of each parameter estimate.

☐ Fit Statistics

Target	Target Label	Fit Statistics	Statistics Label	Train	Validation
Fraudulent_Claim	Fraudulent_Claim	_ASE_	Average Squared Error	0.054254	0.055189
Fraudulent_Claim	Fraudulent_Claim	_DIV_	Divisor for ASE	5994	4002
Fraudulent_Claim	Fraudulent_Claim	_MAX_	Maximum Absolute Error	0.984818	0.986253
Fraudulent_Claim	Fraudulent_Claim	_NOBS_	Sum of Frequencies	2997	2001
Fraudulent_Claim	Fraudulent_Claim	_RASE_	Root Average Squared Error	0.232925	0.234922
Fraudulent_Claim	Fraudulent_Claim	_SSE_	Sum of Squared Errors	325.1987	220.8644
Fraudulent_Claim	Fraudulent_Claim	_DISF_	Frequency of Classified Cases	2997	2001
Fraudulent_Claim	Fraudulent_Claim	_MISC_	Misclassification Rate	0.061061	0.061969
Fraudulent_Claim	Fraudulent_Claim	_WRONG_	Number of Wrong Classifications	183	124

Fig. 4.18 Partial least squares fit statistics

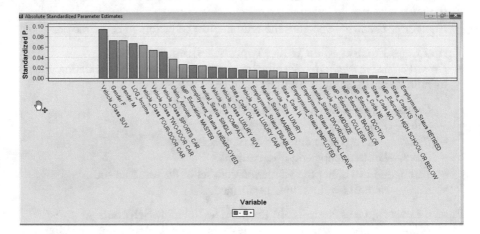

Fig. 4.19 Partial least squares absolute standardized parameter estimate graph

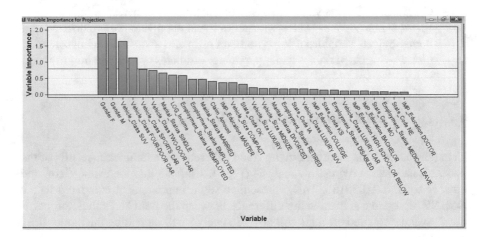

Fig. 4.20 Partial least squares variable importance graph

In addition to the parameter estimates, the partial least squares node results include a variable importance graph (Fig. 4.20). A variable importance score ≥1 is an indication that the input variable has a stronger predictive influence on the target

Fig. 4.21 Partial least squares cumulative lift graph

Fig. 4.22 Least angle
regression node

variable. If the variable importance score is <1, then it has a weaker predictive influence and should be removed from the analysis.

The partial least squares node includes the cumulative lift chart (Fig. 4.21). Note that the lift remains above 2.0 out to a depth of 40% consistent with the results of the DMINE regression. We should continue to consider this model until a better model is developed.

4.12   Least Angle Regression in SAS Enterprise Miner™

A least angle regression (LARS) node (Fig. 4.22) provides an algorithm that is similar to forward stepwise regression that is useful for analyzing data sets that have a large number of input variables. Instead of adding a single variable at a time, it adds a single model parameter at a time, producing a series of regression models. This node also supports least absolute shrinkage and selection operator (LASSO) regression. LASSO regression is designed to use a subset of covariates, rather than all of the covariates for the purpose of creating a more accurate model.

As was the case with partial least squares, the LARS node excludes observations that contain missing values in the target variable. Missing values in a categorical input variable are treated as a separate category.

Fig. 4.23 Partial least squares properties

Property	Value
General	
Node ID	LARS
Imported Data	
Exported Data	
Notes	
Train	
Variables	
Modeling Techniques	
Use Class Inputs	Yes
Intercept	Yes
Variable Selection Method	LAR
Model Selection Criterion	SBC
Path Stopping Criterion	Maximum Steps
Maximum Steps	200
Cross Validation Options	
Cross Validation	Random
CV Fold	5
Seed	12345
Reports	
Details	Summary
Score	
Excluded Variables	Reject
Status	
Create Time	9/29/18 11:21 AM
Run ID	6b9cd21a-9130-46fd-8aef-
Last Error	
Last Status	Complete
Last Run Time	9/29/18 1:18 PM
Run Duration	0 Hr. 0 Min. 41.59 Sec.
Grid Host	
User-Added Node	No

4.12.1 Least Angle Regression Properties

Let's examine the key properties of the least angle regression node (Fig. 4.23).

LARs Node Train Properties: Modeling Techniques
- *Variable Selection Method*—Specifies the variable selection technique that will be used; options include:
    - *LASSO*—Uses the LASSO method for adding and deleting parameters based on the sum of the absolute regression coefficients.
    - *LAR*—Uses the least angle regression method which adds parameter estimates one at a time.
    - *Adaptive LASSO*—Uses the adaptive LASSO method for adding and deleting parameters based on the sum of the absolute regression coefficients constrained by an adaptive weight for penalizing the coefficients.

- *None*—The ordinary least squares regression model is used with no variable selection.

- *Model Selection Criterion*—Specifies the statistical method to choose the optimal model; options include:

  - *SBC*—Schwarz Bayesian information criterion
  - *BIC*—Bayesian information criterion
  - *AIC*—Akaike's information criterion
  - *AICC*—Corrected Akaike's information criterion
  - *CP*—Mallow *C(p)* statistic
  - *Validation*—Uses the average squared error for the validation data
  - *Cross Validation*—Uses the predicted residual sum of square with k-fold cross validation (SAS Enterprise Miner Reference Help 2018).

4.12.2   Least Angle Regression Results

Using the automobile insurance claim fraud data set, let's examine the results of the least angle regression analysis (Fig. 4.24). Note that the average squared error is equal to 0.055608. This is slightly higher than the partial least squares example (which was 0.055189). If no other adjustments were made to further optimize this model, we would conclude that the partial least squares model provided a slightly better fit for this example, and therefore it is the model we would consider for subsequent analysis.

The parameter estimate graph (Fig. 4.25) shows only three parameters that were considered in the final regression model. These include Gender_F (female), Vehicle_Class_SUV, and Vehicle_Class_Sports_Car. This indicates the final model is a rather simplistic model; it's not surprising that it provides a weaker predictive capability.

The least angle regression node provides a graph of the adjusted R-squared value to the iterations used to develop the final regression model (Fig. 4.26). The third iteration was selected as the final model. The adjusted R-squared value is 0.35 which further indicates a rather weak predictive model.

Fit Statistics					
Target	Target Label	Fit Statistics	Statistics Label	Train	Validation
Fraudulent_Claim	Fraudulent_Claim	_ASE_	Average Squared Error	0.054701	0.055608
Fraudulent_Claim	Fraudulent_Claim	_DIV_	Divisor for ASE	5994	4002
Fraudulent_Claim	Fraudulent_Claim	_MAX_	Maximum Absolute Error	0.974045	0.975189
Fraudulent_Claim	Fraudulent_Claim	_NOBS_	Sum of Frequencies	2997	2001
Fraudulent_Claim	Fraudulent_Claim	_RASE_	Root Average Squared Er...	0.233883	0.235814
Fraudulent_Claim	Fraudulent_Claim	_SSE_	Sum of Squared Errors	327.8804	222.5434
Fraudulent_Claim	Fraudulent_Claim	_DISF_	Frequency of Classified ...	2997	2001
Fraudulent_Claim	Fraudulent_Claim	_MISC_	Misclassification Rate	0.061061	0.061969
Fraudulent_Claim	Fraudulent_Claim	_WRONG_	Number of Wrong Classif...	183	124

Fig. 4.24   Partial least squares fit statistics

Fig. 4.25 Least angle regression parameter estimate graph

Fig. 4.26 Least angle regression adjusted R-squared graph

Finally, when we compare the cumulative lift of the least angles' regression model (Fig. 4.27) to the partial least squares model, we see that its lift over the first 40% of the data is also weaker than the cumulative lift of the partial least squares model.

The LARS node provides an alternative method to generate a regression model. In this example, we have evaluated several outputs which demonstrate that this model did not produce a better fit than the partial least squares model. The example illustrates that the regression techniques do not all yield the same results. Further, it may be important in the analysis of a business problem to develop multiple models to determine which model provides the best fit for the problem.

Fig. 4.27 Least angle regression cumulative lift graph

4.13   Other Forms of Regression

Regression remains one of the most popular predictive analytics techniques in part because it is so well known. This includes all of the techniques covered thus far that are directly supported via nodes within SAS Enterprise Miner™ as well as several other types that are not directly supported (Table 4.1).

Ridge regression can be useful when you have a large number of variables but a low ratio on the number of observations to the variables, as well as high multicollinearity. It uses a process to overcome overfitting called regularization by adding a constraint on the sum of the squares of the error coefficient.

Elastic net regression is a variation of lasso and ridge regression that uses a regularization optimization function for linear regression (Hans 2011). It generally provides an improvement when dealing with highly correlated input variables.

Quantile regression is used when there is high skewness, outliers, and heteroscedasticity exists (i.e., conditions for linear regression are not met). Quantile regression estimates the median of the target variable (Dye 2000). Percentiles can be predicted based upon the distribution of the input variables.

Support vector regression can be used for both linear and nonlinear data. It provides the flexibility to predict how much error is acceptable and utilizes nonlinear functions to produce a line that fits the data.

Ordinal regression is used to predict ranked variables (i.e., the target variable is ordinal). The input variables may be categorical or continuous. It can be useful when evaluating Likert scale data because it overcomes the linear regression assumption that the differences between variables are equal.

Poisson regression, *quasi-Poisson regression*, and *negative binomial regression* are used when the target variable has a Poisson distribution, does not contain a decimal, and is not negative. For Poisson regression, it assumes that the variance is equal to the mean. It is sometimes referred to as a log-linear model. Quasi-Poisson

Table 4.1 Types of regression

Regression type	Node within SAS Enterprise Miner™
Linear	Yes
Logistic	Yes
Polynomial	Yes
Partial least squares	Yes
Least angle	Yes
Lasso	Yes
Principal components	Yes
Ridge	No
Elastic net	No
Quantile	No
Support vector	No
Ordinal	No
Poisson	No
Quasi-Poisson	No
Negative binomial	No
Cox	No
Tobit	No

regression uses a linear function of the mean. Negative binomial regression uses a quadratic function of the mean to calculate the variance.

Cox regression (also referred to as proportional hazards regression) is an alternative to survival analysis for analyzing time series (i.e., estimating the amount of time it takes to reach a certain event). It assumes the effects of the predictor variables are constant over time and measures the probability of the event occurring (i.e., the probability must be in the range 0 to 1) (LaMorte 2016).

Tobit regression is used to estimate linear relationships when censoring exists in the target variable (i.e., observations exist for all input variables, but we only know the true value of the target variable for a subset of the observations).

4.14   Chapter Summary

Regression analysis is a popular predictive modeling technique. Some examples where organizations use regression analysis are to forecast future opportunities and risks, predict demand of a product, estimate the number of shoppers in a location, estimate the number of claims made by a policyholder, optimize business processes, analyze wait times at call centers and the number of complaints, optimize marketing campaigns, and detect and prevent fraud. The use of regression analysis provides scientific-based analysis that can lead organizations to make more informed and accurate decisions.

There are numerous regression analysis techniques. Regression models involve the following parameters (input/predictor variables (X)), target (dependent) variable (Y), and unknown parameters (β). There are numerous existing forms of regression techniques used to make predictions. Several common techniques are linear regression, logistic regression, ordinary least squares (OLS), and partial least squares (PLS). Determination on which model to use depends on three factors—the number of independent variables, the dependent variable, and the shape of the regression line. Regression techniques aim to fit a set of data points to a line or a curve such that the distance between the data points from the line or curve is minimized. This is referred to the line of best fit.

Popular linear regression techniques include ordinary least squares, simple linear regression, and multiple linear regression. Five assumptions that must be validated for a model to generate suitable results include linear and additive relationship between the target and input variables, multicollinearity does not exist, variance is constant (heteroskedasticity does not exist), no correlation between the residual (error) terms, and the error terms are normally distributed.

It is essential that a relationship between the variables is determined prior to fitting a linear model to the collected data. Noteworthy is that just because a relationship exists does not mean that one variable causes the other, rather some significant association between the two variables exists. Two common methods to determine the strength of the relationship between two variables are a scatter plot graph and the (Pearson) correlation coefficient (r) measure. If no association between the target and input variable exists, then fitting a linear regression model to the collected data will not provide a suitable model. Popular metrics used to evaluate the strength of the regression line include coefficient of determination (R^2), adjusted R^2, and p-value. If the model contains more than one input variable, adjusted R^2 or p-value should be used. The R^2 and adjusted R^2 (integrates the model's degrees of freedom) provide the percentage variation in the target (Y) variable explained by the input (Y) variables. The range is from 0% to 100%. Typically, the higher the percentage, the more robust the model is. The p-value (probability value) is used to determine the significance of the model results when applying hypothesis testing. The p-value is a number between 0 and 1. Generally, a p-value ≤ 0.05 (considered a small p-value) indicates strong evidence against the null hypothesis (i.e., the assumption that is attempting to be tested); therefore the null hypothesis would be rejected, thus concluding that a linear relationship does exist.

Careful selection of the input (predictor) variables improves model accuracy, avoids overfitted models, reduces model complexity, generates easier interpretation of the model, and trains the model faster. Popular methods useful in input variable selection include clustering, principal component analysis, and sequential methods (forward selection, backward selection, and stepwise selection). The three sequential methods should produce the same models if there is very little correlation among the input variables and no outlier problems. A limitation exists with small data sets.

Principal component regression is an extension of principal component analysis. Principal component regression takes the untransformed target variable and regresses it on the subset of the transformed input variables (principal components) that were not removed. This is possible since the newly transformed input variables are independent (uncorrelated) variables.

Partial least squares (PLS) is a flexible regression technique that has features of principal component analysis and extends multiple linear regression. PLS is best used when the data set contains fewer observations than input variables and high collinearity exists. PLS can also be used as an exploratory analysis tool to identify input variables to include in the model and outliers. Like multiple linear regression, PLS' overarching goal is to create a linear predictive model.

The regression models discussed so far have a numeric (continuous) target variable, and the data set is normally distributed. Generalized linear models are an extension of traditional linear models and are used when the target variable is discrete, and the error terms do not follow a normal distribution. Logistic regression belongs in this category of regression models and is one of the most commonly used techniques for discrete data analysis. Logistic regression is used when examining the relationship between a categorical (class) target variable and one or more input variables. The target variable can be either binary, multinomial, or ordinal. The input variables can be either continuous (numeric) or categorical. Logistic regression provides the likelihood or probability a specific outcome/event will occur. Logistic regression predicts the value of the outcome. Logistic regression violates the two linear regression assumptions. First, residuals are not normally distributed, and second constant variance of the error terms does not exist. Assumptions of logistic regression include adequate sample size, multicollinearity does not exist, errors need to be independent but do not have to be normally distributed, and there are no outliers. The nature of the categorical (class) target variable determines which of the three types of logistic regression will be used—binary logistic regression, multi-nominal logistic regression, and ordinal logistic regression.

In closing, regression methods are popular predictive analysis techniques and easy to implement and understand. There are numerous forms of regression. Some are very widely utilized; others are specialized. Some of these are directly supported by software such as SAS Enterprise Miner™. Others require statistical programming languages such as R, Python, or SAS. In the upcoming chapters, decision tree analysis, neural networks, and several other techniques will be examined and built upon regression techniques.

Discussion Questions
1. Describe a business problem where it would be appropriate to use regression for analysis.
2. Describe conditions which may cause significant differences between the backward, forward, and stepwise methods.
3. There are many different types of regression. Research one other type, not covered in this chapter. Discuss how it compares to linear or logistic regression.
4. Explain what is meant by underfitting and overfitting, and describe an example of each.
5. Present a checklist for all of the steps that should be performed to complete a multiple linear or logistic regression.

References

Dye S (2000, February 12) Quantile regression. towardsdatascience.com/quantile-regression-
    ff2343c4a03
Hans C (2011) Elastic net regression modeling with the orthant normal prior. J Am Stat Assoc
    106(496):1383–1393
LaMorte W (2016, June 3) Survival analysis. sphweb.bumc.bu.edu/otlt/mph-modules/bs/bs704_
    survival/BS704_Survival6.html
SAS Enterprise Miner Reference Help (2018) SAS Enterprise Miner Version 14.2 Software.
    SAS, Cary

Chapter 5
The Second of the Big 3: Decision Trees

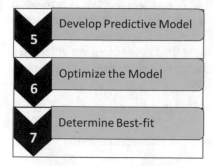

Chapter Learning Objectives
1. Describe decision trees.
2. Create and utilize decision trees.
3. Evaluate the key properties that define a decision tree.
4. Evaluate the results of a decision tree output.

Decision trees are an important predictive modeling tool, not because of their complexity but because of their simplicity. They are often used to be able to provide an easy method to determine which input variables have an important impact on a target variable. In this chapter, decision trees are defined and then demonstrated to show how they can be used as an important predictive modeling tool. Both classification and regression decision trees will be considered. Decision trees can be system generated or built interactively; both will be demonstrated. Focus will be given to interpreting the output of a decision tree. Decision trees are useful when it is important to understand and even control which variables impact a target.

5.1 What Is a Decision Tree?

Decision trees are one of the most widely used predictive and descriptive analytic tools. One reason for their popularity is that they are easy to create, and the output is easy to understand. They are particularly useful in situations where you want to know specifically how you arrived at an outcome. Decision trees require at least one target variable which can be continuous or categorical. Decision trees use algorithms to determine splits within variables that create branches. This forms a

© The Author(s), under exclusive license to Springer Nature
Switzerland AG 2022
R. V. McCarthy et al., *Applying Predictive Analytics*,
https://doi.org/10.1007/978-3-030-83070-0_5

tree-like structure. The algorithms create a series of *if-then-elese* rules that split the data into successively smaller segments. Each rule evaluates the value of a single variable and based upon its value splits it into one of two or more segments. The segments are called *nodes*. If a node does not result in a subsequent split (i.e., it has no successor nodes), it is referred to as a *leaf*. The first node contains all of the data and is referred to as the *root node*. A node with all of its successor nodes is referred to as a *branch* of the decision tree. The taller and wider the decision tree is, the more data splits that will have occurred. Decision trees can be a great place to begin predictive modeling to gain a deeper understanding of the impact input variables have on a target variable; however, they rarely produce the best-fit model. A decision tree can be used to perform one of the following tasks:

1. Classify observations based upon a target that is binary, ordinal, or nominal.
2. Predict an outcome for an interval target.
3. Predict a decision when you identify the specific decision alternatives (Scientific Books 2016).

Decision trees are frequently used for market or customer segmentation. Consider the following simple example of a decision tree (Fig. 5.1). A bank wants to make a mortgage decision based upon credit rating. The root node is the mortgage decision. The first branch is the credit rating. The node for a low credit rating is a leaf because there is no subsequent node from that branch. The node for a high credit rating is a branch as it has subsequently been split on the income variable creating two additional leaves.

In addition to providing a means to describe the rules by which data can be evaluated, decision trees can be useful because they can handle missing data values. Missing values are treated as a category of data when applying split rules. If the missing data prohibits the splitting of the data, then a surrogate rule can be applied. A surrogate rule is an alternative rule used to handle missing values. Split rules can be evaluated either by utilizing a statistical significance test (e.g., Chi-square or

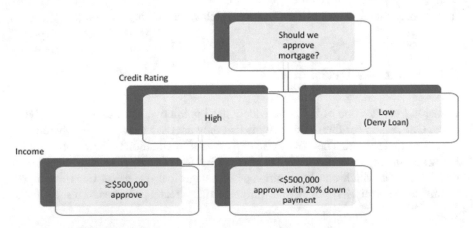

Fig. 5.1 Decision tree example

F-test) or based on a reduction in entropy, variance, or Gini impurity. The statistical significance test is used to determine which values to combine. The strength of the significance determines whether the values have a strong relationship to the target and therefore should be combined, or if the relationship is weak, they should not be combined. DeVille and Neville (2013) report the following guideline for determining the strength of the relationship (Table 5.1).

In SAS Enterprise Miner™, a decision tree can be system generated, or an interactive tree can be built.

5.2   Creating a Decision Tree

To create a decision tree, a default algorithm is used to split the data. The algorithm begins with a *split search*. The split search begins by selecting an input variable to be split (i.e., create subsets). If the input variable has an interval data type, each unique value within the test partition serves as a potential split. If the input variable has a categorical data type, the average value of the target within each category is computed. For the selected input variable, two groups are created. The left branch includes all the cases less than the split value; the right branch contains all the cases greater than the split value. When combined with a binary target variable, a 2X2 matrix is created (Fig. 5.2).

The counts within each column are quantified using a Pearson's Chi-square statistic. A good split can be determined by a large difference in the outcome proportions between the left and right branches. For large data sets, the quality of a split is reported by logworth (logworth = log(Chi-squared p-value). The highest logworth indicates the best split for an input variable. This process is repeated for every input variable in the training data set. Input variables with an adjusted logworth less than

Table 5.1   Confidence values

Confidence	Strength of the relationship
0.001	Extremely good
0.01	Good
0.05	Pretty good
0.10	Not so good
0.15	Extremely weak

Fig. 5.2 Example of a decision matrix

Target	Left	Right
0	53	47
1	41	59

the threshold (default of 0.70) are excluded from the resulting decision tree (SAS 2015).

After the nodes are split, they can then be evaluated to determine if they can be split further. This process continues until a stop condition is reached. A stop condition is reached when one of the following occurs:

1. The maximum depth of the tree has been reached (i.e., the number of nodes between the root node and the given node).
2. The nodes can no longer be split in a meaningful way (i.e., the threshold worth).
3. The nodes contain too few observations to be meaningful.

5.3   Classification and Regression Trees (CART)

Decision trees are categorized as classification tree models or regression tree models. This is determined by the data type of the target variable. Categorical target variables result in a classification tree. Interval target variables result in a regression tree. The goal in both cases is to build a model that predicts a target variable based upon the input variables. The major difference between classification and regression trees is the test statistics used to evaluate the splits. For a regression tree, the test statistic used has an F-distribution (i.e., ANOVA), and the assessment measure is the average squared error (SAS 2015).

Within a classification decision tree, three of the most widely used clustering algorithms to produce and evaluate the splits are entropy, Gini impurity, and Chi-square (de Ville and Neville 2014). Entropy is a measure of the variability of the splits in the leaves. The entropy of a split is calculated by first calculating the entropy of all of the leaves contained within that split and then summing them together. The entropy of the outcome of a leaf is $-Log^2$ (P_i), where p is the relative frequency of the categorical variable within the node.

The *Gini index*, which should not be confused with the Gini coefficient (described in Chap. 7), assesses the quality of the data (i.e., it is a measure of the impurity of the node). The index was proposed by Breiman et al. (1984) and is calculated as

$$\text{Gini index} = 1 - \sum p^2$$

When this index is zero, the node is considered as pure. Alternatively, when it comes close to the value of 1, the node is considered impure. The splitting criterion uses the data that has an index that most closely approaches zero.

The Chi-square clustering process is built upon a Chi-square or t-test of significance. Clusters of input variables are formed by evaluating their relationship to the target variable and grouping those that are similar. Two input variables are considered similar if their difference between their value and the target variable is not considered to be statistically significant. A t-test is used to test for statistical significance of an interval target variable. A Chi-square test is used to test for statistical significance of a categorical target.

Building a decision tree has three main goals:

1. Decrease the entropy (unpredictability of the target variable—also referred to as the information gain).
2. Ensure that data set is reliable and consistent.
3. Have the smallest number of nodes.

Decision trees can be helpful in creating segments of a larger group. For example, many retail organizations have loyalty cards that are capable of collecting large amounts of data about customer spending patterns. A decision tree can help identify the characteristics of sub-groups of customers that have similar spending patterns.

5.4   Data Partitions and Decision Trees

As discussed previously, when predictive models are built, the data is partitioned so that the model can be constructed and subsequently tested. In SAS Enterprise Miner™, there should always be a training data set and a validation data set. Optionally, if the data set is large enough, a test data set may be created. If present it will provide an independent test of the decision tree model.

The training data set is used to generate the initial decision tree and to develop the splitting rules to assign data to each of the nodes. The training data set assigns each node to a target level to be able to determine a decision. The proportion of cases within the training data set that are assigned to each node (referred to as the posterior probabilities) is calculated. As is the case with other predictive modeling techniques, the general rule of thumb is the smaller the data set, the larger the percentage that should be allocated to the training data set. In general, the posterior possibilities of the node will be more reliable from a larger training data set.

The training data set will produce an initial or *maximal tree*. The maximal tree is the whole tree or largest possible tree. However, the optimal tree is the smallest tree that yields the best possible fit. Optimizing a decision tree is referred to as *pruning* the tree. The validation data set is used to prune a decision tree. Pruning a tree is a process of evaluating the results of subtrees by removing branches from the maximal tree. Once the subtrees are built, SAS Enterprise Miner™ provides three possible methods to determine which subtree to use. These are:

1. Assessment (i.e., the best assessment value)
2. Largest (i.e., the subtree with the most leaves)
3. N (i.e., the subtree that contains the number of leaves indicated by the user)

To illustrate this process, let's examine a split on marital status using a subset of data from the claim fraud case (Table 5.2).

There are three possible categories for marital status (Fig. 5.3): single, divorced, and married.

Table 5.2 Sample training data set—fraudulent claims

Claimant_Number	Gender	Marital_Status	Vehicle_Size	Fraudulent_Claim
1	F	Divorced	Compact	No
2	M	Married	Midsize	No
3	F	Single	Compact	Yes
4	M	Married	Midsize	No
5	F	Divorced	Luxury	No
6	M	Divorced	Luxury	No
7	M	Single	Compact	Yes
8	M	Married	Midsize	No
9	F	Single	Compact	No
10	M	Married	Luxury	No
11	F	Single	Compact	No
12	M	Single	Luxury	No
13	F	Single	Midsize	Yes
14	M	Single	Midsize	No
15	M	Married	Midsize	No
16	M	Married	Compact	Yes
17	M	Divorced	Luxury	No
18	M	Married	Midsize	No
19	M	Single	Compact	Yes
20	F	Married	Midsize	No

Fig. 5.3 Marital status split

In this example, there were eight (8) observations where the claimant is single, four (4) observations where the claimant is divorced, and eight (8) observations where the claimant is married. For the divorced cases, there are no fraudulent claims. The branch is considered pure (i.e., there is no need to split it further; all of the target values are the same). No further pruning of the divorced branch would occur. The single and married branches could be split further based upon other input variables (in this case gender and vehicle size).

5.5   Creating a Decision Tree Using SAS Enterprise Miner™

Let's examine the use of a decision tree as a predictive analytics tool. Two decision trees will be applied to the claim fraud case, and we will examine the results to determine which decision tree provides the best predictive capability.

The first decision tree that has to be utilized will be a generated tree (Fig. 5.4). The Decision Tree node is located within the Model tab of the SEMMA model.

Now let's examine the properties of this decision tree (Figs. 5.5 and 5.6).

There are several properties that are essential to controlling the decision tree that will be generated. The following summarizes critical properties of the *Decision Tree* node:

Train Properties
- *Variables*—Specifies the properties for each variable. The variable properties can be changed for the decision tree only; however, in most cases, it is advisable to change them on the data source so that all other nodes use the same properties. Select the ▦ button to open the variables table.
- *Interactive*—This is used to create a user-defined customized decision tree (an example is described below).
- *Frozen Tree*—When set to Yes, it prevents the maximal decision tree from being changed by other settings. The default is No.

Splitting Rule Properties
- *Interval Target Criterion*—Identifies one of two methods used for splitting when the target is an interval variable.
  - *ProbF*—Utilizes the *p*-value of the F-test (i.e., log worth) that is associated with the node variance.
  - *Variance*—Utilizes the reduction in the square error from the mean of the node.
- *Nominal Target Criterion*—Identifies one of three methods used for splitting when the target is a nominal variable.
  - *ProbChisq*—Utilizes the *p*-value of the Pearson Chi-square statistic for the target versus the branch node.

Fig. 5.4 Decision Tree node

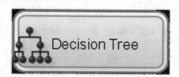

Fig. 5.5 Decision tree
properties

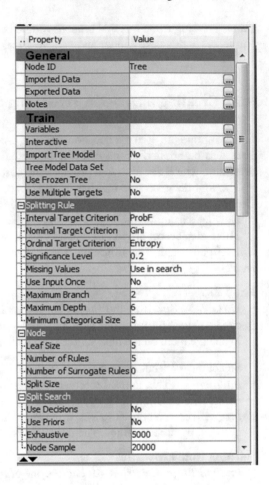

- *Entropy*—Reduction in the entropy measure.
- *Gini*—Reduction in the Gini index.

• *Ordinal Target Criterion*—Identifies one of two methods used for splitting when the target is an ordinal variable.

- *Entropy*—Reduction in the entropy measure, adjusted with ordinal distances.
- *Gini*—Reduction in the Gini index, adjusted with ordinal distances.

• *Significance Level*—When a *ProbChisq* or *ProbF* is chosen, this defines the maximum appropriate *p*-value for the worth of a candidate splitting rule. Acceptable significances are numbers greater than 0 and less than or equal to 1. The default is 0.2; the lower the number, the more restrictive the significance level is. Setting this close to 1 may result in a model that is too inclusive and thus not useful for predictive analysis.

Fig. 5.6 Additional
decision tree properties

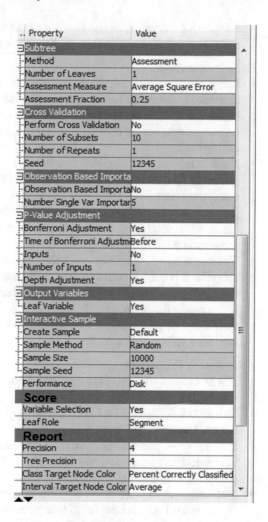

.. Property	Value
⊟Subtree	
Method	Assessment
Number of Leaves	1
Assessment Measure	Average Square Error
Assessment Fraction	0.25
⊟Cross Validation	
Perform Cross Validation	No
Number of Subsets	10
Number of Repeats	1
Seed	12345
⊟Observation Based Importa	
Observation Based Importa	No
Number Single Var Importar	5
⊟P-Value Adjustment	
Bonferroni Adjustment	Yes
Time of Bonferroni Adjustm	Before
Inputs	No
Number of Inputs	1
Depth Adjustment	Yes
⊟Output Variables	
Leaf Variable	Yes
⊟Interactive Sample	
Create Sample	Default
Sample Method	Random
Sample Size	10000
Sample Seed	12345
Performance	Disk
Score	
Variable Selection	Yes
Leaf Role	Segment
Report	
Precision	4
Tree Precision	4
Class Target Node Color	Percent Correctly Classified
Interval Target Node Color	Average

- *Missing Values*—Specifies how splitting rules handle missing values. The three
  possible choices are use in search (default), largest branch, and most correlated branch.
- *Use Input Once*—Specifies whether a splitting variable may be evaluated only
  once (the default) or repetitively in splitting rules that relate to descendant nodes.
- *Maximum Branch*—Specifies the maximum number of branches. Two is the
  default, which generates binary branches. It is common to use four or five
  branches. However, consideration should be given to the data to determine an
  appropriate value.
- *Maximum Depth*—Specifies the largest number of groups of nodes in the decision tree. The default is 6; however the range can be from 1 to 150. Very large
  trees can result in too many splits and may be impractical to explain.

- *Minimum Categorical Size*—Specifies the smallest number of training observations. The default value is five training observations.

Node Properties
- *Leaf Size*—Specifies the minimum number of training observations. The splitting rules try to create leaves that are as alike as possible within the same branch but different when compared to leaves of other branches at the same level. This property therefore specifies a minimum variability within the tree. The default setting is five.
- *Number of Rules*—A decision tree uses only one rule to determine the split of a node, but others may have been attempted before the final split. This property determines how many rules will be saved for comparison. A splitting rule statistic is computed. LOGWORTH is used for ProbChisq or ProbF statistics; WORTH (variance) is used for entropy or Gini. The default value for the Number of Rules property is 5.
- *Number of Surrogate Rules*—A surrogate rule is a back-up to the primary splitting rule. It is used to split non-leaf nodes and invoked when the primary splitting rule relies on a variable with a missing value. The default value for the Number of Surrogate Rules property is zero.
- *Split Size*—Specifies the minimum number of training observations to be considered for a split within a node.

Split Search Properties
- *Use Decisions*—When set to *Yes*, it will apply the decision criteria during a split search. The default is No.
- *Use Priors*—When set to *Yes*, it will apply prior probabilities during a split search. The default is No.
- *Exhaustive*—Specifies the maximum number of candidate splits. Allowable values are integers between 0 and 2,000,000,000. This is a consideration when evaluating big data data sets. The default setting for the Exhaustive property is 5000.
- *Node Sample*—Specifies the maximum sample size to use within a node split. Allowable values are integers greater than or equal to 2. The default value for the Node Sample property is 20,000.

Additional properties are shown in Fig. 5.6.
The key properties include:

Subtree Properties
- *Method*—Specifies how a subtree will be constructed. The options include:

    - *Assessment*—This is the default. The smallest subtree with the best assessment value will be utilized. The validation data set is used.
    - *Largest*—The full tree is chosen.
    - *N*—The largest subtree containing at least N leaves is chosen.

- *Number of Leaves*—When the Method property is set to N, this specifies the largest number of leaves that will be contained within a subtree.

- *Assessment Measure*—Specifies the tree selection method (i.e., the criteria to determine the best tree) based on the validation data. If there is no validation data set, the training data set will be used.

The assessment measurements include:

- – *Decision* (default setting)—If a profit or loss matrix is defined, this chooses the tree that has the largest average profit and lowest average loss. If the target is categorical, the measure is set to Misclassification.
- – *Average Squared Error*—Chooses the tree that has the lowest average squared error.
- – *Misclassification*—Chooses the tree that has the smallest misclassification rate.
- – *Lift*—Assesses the tree based on the posterior probability (for categorical or ordinal targets) or predicted target value (for interval targets) projection of the top *n%* of the ranked observations. The *n%* is set with the *Assessment Fraction property*. The default value for the Assessment Fraction property is 0.25.

P-Value Adjustment Properties
- *Bonferroni Adjustment*—This is a Yes/No value that indicates if a Bonferroni adjustment to the *p*-values is performed. The default setting is *Yes*. Bonferroni adjustments are a conservative adjustment to account for the number of tests that were used. Bonferroni adjustment provides a method to place tests that used hundreds of input variables on the same level as those that only used a small number of input variables. When utilized in the absence of any other criteria and when the time of the adjustment is set to before, the input variable with the highest Bonferroni adjustment will be the one selected from the split.
- *Time of Bonferroni Adjustment*—Indicates whether the Bonferroni adjustment should take place *Before* or *After* the split is chosen. The default setting is *Before*. The Time of Bonferroni Adjustment property is ignored if the Bonferroni Adjustment property is set to *No*.
- *Inputs*—This property is typically used when you have many correlated input variables to reduce the risk that one of the correlated input variables will generate too many type I errors (DeVille and Neville 2013). When Inputs is set to *Yes*, the *Number of Inputs* property should be set to identify the number of input variables that are not correlated.

Once the decision tree is executed, the primary output to consider is of course the decision tree itself (Fig. 5.7). Note the color differences in the nodes. The darker the color, the purer the node is. A white node indicates a weaker result. In all cases, the node contains the probabilities of each outcome of the target variable for both the test and validation data sets. For example, looking at the two-door car leaf, the probability or percentage of two-door cars being fraudulent is 8.75%. The thickness of the lines indicates the volume of observations that passed through that path. The thicker the line, the greater the number of observations is.

This decision tree considered four input variables: vehicle class, gender, transformed income, and employment status. It produced a split on only two nodes within

Fig. 5.7 Decision tree diagram results

Variable Name	Label	Number of Splitting Rules	Importance	Validation Importance	Ratio of Validation to Training Importance
Vehicle_Class	Vehicle_Class	1	1.0000	1.0000	1.0000
Gender	Gender	1	0.9227	0.8727	0.9458
Employment_Status	Employment_Status	1	0.6438	0.6860	1.0656
LOG_Income	Transformed: Income	1	0.2572	0.1684	0.6549
Marital_Status	Marital_Status	0	0.0000	0.0000	
IMP_Outstanding_Balance	Imputed: Outstanding_Balance	0	0.0000	0.0000	
State_Code	State_Code	0	0.0000	0.0000	
IMP_Education	Imputed: Education	0	0.0000	0.0000	
Vehicle_Size	Vehicle_Size	0	0.0000	0.0000	

Fig. 5.8 Variable importance table

vehicle class and again on only one node within gender. Note that the splits within vehicle class are not on the same variable. The variable importance table (Fig. 5.8) provides a listing of the input variables that were used to produce the decision tree along with the number of splits that occurred within those variables. Within the table, important is the observed value of the variable importance statistic for the training data set. The importance statistics evaluates the impact of a variable over the entire fitted tree. The variable, vehicle_class, impacts the entire tree, while those variables lower in the tree, gender, for example, impacts a smaller number of leaves or branches. Validation importance is the observed value of the variable importance statistics for the validation statistic. The Ratio of Validation to Training Importance is the ratio of the validation importance statistics to the training importance statistics. If the ratio is small, it indicates a variable that was used in overly optimistic splitting rules (i.e., a variable whose actual explanatory contribution is lower than its estimated contribution). When optimizing the tree, this may be a candidate for pruning.

The fit statistics (Fig. 5.9) indicate the decision tree resulted in an average squared error of 0.055307. The fit statistics are used to compare decision trees to

| Fit Statistics | | | | | |
Target	Target Label	Fit Statistics	Statistics Label	Train	Validation
Fraudulent_Claim	Fraudulent_Claim	_NOBS_	Sum of Frequencies	2997	2002
Fraudulent_Claim	Fraudulent_Claim	_MISC_	Misclassification Rate	0.061061	0.061938
Fraudulent_Claim	Fraudulent_Claim	_MAX_	Maximum Absolute Error	0.997886	1
Fraudulent_Claim	Fraudulent_Claim	_SSE_	Sum of Squared Errors	325.8853	221.4477
Fraudulent_Claim	Fraudulent_Claim	_ASE_	Average Squared Error	0.054369	0.055307
Fraudulent_Claim	Fraudulent_Claim	_RASE_	Root Average Squared Error	0.233171	0.235174
Fraudulent_Claim	Fraudulent_Claim	_DIV_	Divisor for ASE	5994	4004
Fraudulent_Claim	Fraudulent_Claim	_DFT_	Total Degrees of Freedom	2997	

Fig. 5.9 Decision tree fit statistics

Fig. 5.10 Leaf Statistics

other predictive models. For example, in the logistic regression model in Chap. 4 that used the backward method, the average squared error was 0.056825. We can conclude that this decision tree is a slightly better predictive model (because it has a lower average squared error).

The Leaf Statistics (Fig. 5.10) denotes the number of leaves within the tree, in this case 13. This would also be obtained by counting the leaves within the decision tree. However, for larger more complex trees, this can be cumbersome.

The cumulative lift chart (Fig. 5.11) provides an indication of the performance of the decision tree.

Note that in this example, over the first 25% of the data, the cumulative lift is over 2.1. This provides an indication of the strength of this decision tree. For this example it demonstrates how quickly a model can be built that provides better predictive capability than random guessing, with a graphical display of the decision tree which denotes the contribution of each variable involved in the decision. We were then able to compare this model to the backward regression model to conclude it performs slightly better.

Fig. 5.11   Cumulative lift chart

Fig. 5.12   Subtree assessment plot

5.5.1   *Overfitting*

With any predictive models, there is a risk of overfitting. The model could fit the data so well that it does not generalize (i.e., the model does not work well when applied to a new data set). The subtree assessment plot provides a graphical indication if the model has been overfit (Fig. 5.12).

Note that the subtree assessment plot demonstrates that the average squared error continues to decrease as the number of leaves increases over the relevant range (i.e., the 13 leaves that comprise the tree). If the average squared error beings to increase, the number of leaves should be reduced (i.e., the tree should be pruned).

5.6 Creating an Interactive Decision Tree Using SAS Enterprise Miner™

An interactive decision tree allows you to create your own decision tree, thereby controlling the variables and values of the variables in which splits will occur. One of the advantages of a decision tree is its ability to provide graphical presentation of the input variables and how they split to create an optimal tree. Often the tree that is created will split on values that do not necessarily match how a business evaluates their data. For example, if a variable such as age splits on values such as 23.7, 38.9, and 51.2, these values may be to precise. These values can be adjusted by creating an interactive decision tree. In addition, an interactive decision tree enables the user to include variables for analysis that were not selected within the optimal tree (in the event it was important to the organization to consider that variable).

Once the interactive tree property is selected, the interactive decision tree view is opened (Fig. 5.13) with a decision tree that only contains the root node.

The *Rules Pane* on the left shows the split count from the root node, the node ID for the current node and each predecessor node, the variable utilized for this exact splitting decision, and the split criteria. The *Statistics Pane* on the right provides counts of the observations within each data set, statistics on the percentage of values within each target value, and a predicted result.

Right clicking on the root node enables a split to occur on the root node (Fig. 5.14).

Once split rule is selected, all the variables that could possibly be split are presented for the purpose of selecting the first split (Fig. 5.15).

In this example, a decision tree that splits on IMP_Outstanding_Balance will be created. Once the variable is selected and Edit Rule is chosen, the split criteria can be created (Fig. 5.16). For this example, we can see that four splits took place. In

Fig. 5.13 Interactive decision tree

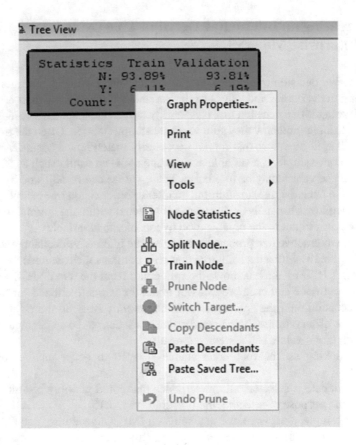

Fig. 5.14 Creating a split for an interactive decision tree

this example however, rather than precise split that occurs from the system-generated decision tree, these splits are based upon values considered to be a more logical separation of the data by the decision-makers based upon the data.

To illustrate how the interactive decision tree can grow, let's add a second split (Fig. 5.17). In this case we will select the node that has a balance between 10,000 and 15,000, splitting on the variable martial_status. This results in a tree that contains six leaves. The example illustrates a user-driven process creating a decision tree based upon variables and splits within those variables that are specifically of interest.

Variable	Variable Description	-Log(p)	Branches
Split Node 1			
Target Variable: Fraudulent_Claim			
Gender	Gender	15.185	2
Vehicle_Class	Vehicle_Class	13.2166	2
Marital_Status	Marital_Status	1.6923	2
LOG_Months_Since_Last...	Transformed: Months_Si...	0.7441	2
Vehicle_Size	Vehicle_Size	0.5077	2
IMP_Outstanding_Balance	Imputed: Outstanding_...	0.4715	2
Employment_Status	Employment_Status	0.4508	2
Income	Income	0.1225	2
IMP_Education	Imputed: Education	0.101	2
State_Code	State_Code	0.0901	2

Edit Rule...

OK   Cancel   Apply   Refresh

Fig. 5.15 Split node selection

Fig. 5.16 Interactive decision tree—first node split

Fig. 5.17   Interactive decision tree—second node split

Target	Target Label	Fit Statistics	Statistics Label	Train	Validation
Fraudulent_Claim	Fraudulent_Claim	_NOBS_	Sum of Frequencies	2997	2002
Fraudulent_Claim	Fraudulent_Claim	_MISC_	Misclassification Rate	0.061061	0.061938
Fraudulent_Claim	Fraudulent_Claim	_MAX_	Maximum Absolute Error	0.969325	0.969325
Fraudulent_Claim	Fraudulent_Claim	_SSE_	Sum of Squared Errors	343.1091	232.5276
Fraudulent_Claim	Fraudulent_Claim	_ASE_	Average Squared Error	0.057242	0.058074
Fraudulent_Claim	Fraudulent_Claim	_RASE_	Root Average Squared Error	0.239253	0.240985
Fraudulent_Claim	Fraudulent_Claim	_DIV_	Divisor for ASE	5994	4004
Fraudulent_Claim	Fraudulent_Claim	_DFT_	Total Degrees of Freedom	2997	.

Fig. 5.18   Interactive decision tree fit statistics

The results of this interactive decision tree yield an average squared error of 0.058074 (Fig. 5.18). When the results of the interactive decision tree are compared to the system-generated decision tree, the system-generated tree is a slightly better fit. Note that the cumulative lift for the interactive decision tree is much worse (Fig. 5.19), with only approximately 15% of the case having a lift great than 1. Generally, the system-generated decision tree will provide a better fit; the primary advantage of the interactive decision tree is the ability to control the values of the node splits.

5.7   Creating a Maximal Decision Tree Using SAS Enterprise Miner™

An interactive decision tree is a process where you consider each variable that will be utilized within a tree and then select the criteria for splitting nodes to form the final tree. An alternative approach is to begin first by creating the maximal tree. The maximal tree is the largest tree structure that can be created based upon the data. Using the maximal tree as a starting point, the decision tree can then be pruned to an optimal tree for use in predictive modeling. To create a maximal tree in SAS Enterprise Miner™, select the interactive tree property from a Decision Tree node.

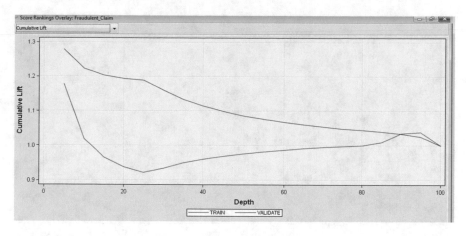

Fig. 5.19 Cumulative lift of the interactive decision tree

Fig. 5.20 Maximal tree

Right-click on the root node, and select *Train Node*. The tree is grown until the stopping rules prohibit it from growing any further.

Using the claim fraud example, the maximal tree (Fig. 5.20) is close to the system-generated tree, using the same variables with differences in the splits.

The maximal tree results in a tree with four branches and eight leaves (the leaves are also shown in the Leaf Statistics, Fig. 5.21).

Three of the leaves contribute very little to the model and are therefore candidates to be pruned. These include the two-door/four-door leaf (in both cases) and the income group between 11.2257 and 11.2593. The cumulative lift of the maximal tree (Fig. 5.22) is slightly lower than the system-generated tree; however it maintains that lift over a larger percentage of cases.

The result of the maximal decision tree yields an average squared error of 0.054817 (Fig. 5.23) which is the best performance of the three trees. Three approaches to creating decision trees have demonstrated how quickly they can be built. In each case, though there are differences in the trees themselves, the results

Fig. 5.21 Leaf Statistics

Fig. 5.22 Cumulative lift of the maximal decision tree

(based on assessment statistics) did not vary greatly. An optimal decision tree could be created using any of these approaches as a starting point.

Decision trees are useful when you have data sets that have a large number of input variables, especially when there are nominal variables. They can be useful for segmenting the nominal variables and ranges that are insignificant and can be easily pruned. Decision trees are a popular technique in marketing analytics because of their ability to provide flexible analysis of market segments and to assist in developing profiles for target marketing. However, they are used in other applications as well. An advantage of their use is that they produce a concise set of clearly identifiable rules that describe how they arrived at their prediction or estimate.

Results - Node: Maximal Tree Diagram: Data Preparation and Analysis

File Edit View Window

Fit Statistics

Target	Target Label	Fit Statistics	Statistics Label	Train	Validation
Fraudulent_Claim	Fraudulent_Claim	_NOBS_	Sum of Frequencies	2997	2002
Fraudulent_Claim	Fraudulent_Claim	_MISC_	Misclassification Rate	0.061061	0.061938
Fraudulent_Claim	Fraudulent_Claim	_MAX_	Maximum Absolute Error	0.944444	1
Fraudulent_Claim	Fraudulent_Claim	_SSE_	Sum of Squared Errors	322.6598	219.4869
Fraudulent_Claim	Fraudulent_Claim	_ASE_	Average Squared Error	0.05383	0.054817
Fraudulent_Claim	Fraudulent_Claim	_RASE_	Root Average Squared Err...	0.232014	0.23413
Fraudulent_Claim	Fraudulent_Claim	_DIV_	Divisor for ASE	5994	4004
Fraudulent_Claim	Fraudulent_Claim	_DFT_	Total Degrees of Freedom	2997	.

Fig. 5.23 Maximal tree fit statistics

5.8 Chapter Summary

Decision trees are a powerful analytic tool because they provide a clear description of each of the input variables that impact a target variable. Decision trees are easy to create and require very little data preparation. A decision tree can also easily handle missing values. A decision tree is created by first beginning with a root node. Using splitting algorithms, the root node is then separated into branches and leaves. Leaves represent an end point (i.e., there are no further splits). A branch is a node that is split further based upon some other variable. This splitting process continues until a stopping condition has occurred, at which time the tree is then fully created.

Decision trees provide a clear set of if-then-else rules that describe segments of the data. Split rules are evaluated by using statistical tests of significance (such as Chi-square or F-tests) or by testing for a reduction in entropy, variance, or Gini impurity. The strength of the significance determines if the values have a strong relationship to the target and therefore will be combined or if the relationship is weak and should not be combined.

The training data set is used to build an initial decision tree. This creates the splitting rules. A maximal, or largest possible, tree can also be produced from the training data set. The validation data set is then used to prune (optimize) the tree.

When creating a decision tree, there are many properties to consider. These define the size of the maximal tree as well as how the splitting rules will be evaluated. Like most predictive modeling techniques, a decision tree must be evaluated to ensure that overfitting or underfitting did not take place. This can be done by analyzing the subtree assessment plot.

Using SAS Enterprise Miner™, a decision tree can be system generated or built interactively. The advantage of an interactive tree is that it provides the user with the ability to control the structure of the tree (i.e., how and where splits will occur). A disadvantage to this approach however is that is rarely results in the optimal decision tree.

When creating a decision tree using SAS Enterprise Miner™, there are several very useful outputs that provide a detailed understanding of the tree and how to use it for variable estimation or prediction. The Leaf Statistics provides an overview of the complexity of the tree. In general, we seek the smallest tree that has the most reliable predictive capability. The cumulative lift chart presents a graphical display of the performance of the decision tree for predictive analysis. The variable importance table provides a list of all the input variables used to create the tree with a measure of the importance of each of these variables.

Though rarely the optimal predictive modeling technique, decision trees remain a highly valuable method to be able to quickly understand complex relationships among input variables and provide a quick and easy method to segment large data sets into more manageable groups.

Discussion Questions
1. Discuss an example where a decision tree can be used to explain the results of a business problem/issue.
2. SAS Enterprise Miner™ supports automatically training and pruning a tree and interactively training a tree with a wide variety of options, based upon the setting of properties for these models. Describe three properties that are part of the Decision Tree node.
3. Explain what is meant by *pruning a tree*.
4. Using the claim fraud example, how can you modify the maximal tree to improve the maximal tree model?
5. Using SAS Enterprise Miner™, create a decision tree to solve a problem. Report your assumptions and findings.

References

Breiman L, Friedman J, Stome C, Olshen R (1984) Classification and regression trees. Taylor & Francis, Boca Raton
De Ville B, Neville (2013) Decision trees for analytics using SAS Enterprise Miner. SAS Institute Inc., Cary
SAS Institute Inc (2015) Getting started with SAS Enterprise Miner 14.1, SAS Documentation. SAS Institute Inc, Cary
Scientific Books (2016) Predictive models to risk analysis with neural networks, regression, and decision trees

Chapter 6
The Third of the Big 3: Neural Networks

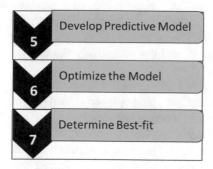

Chapter Learning Objectives
1. Create and utilize neural networks.
2. Differentiate between different neural network architectures.
3. Evaluate the key properties that define and optimize a neural network.
4. Utilize a decision tree to explain a neural network.

One of the most powerful predictive analytics techniques are neural networks. The concept of the neural network is over 50 years old, but it is recent advances in computing speed, memory, and data storage that have enabled their more current widespread use. In this chapter a variety of different neural network architectures will be described. Next an analysis of how to optimize and evaluate neural networks will be presented, followed by using a decision tree to show how to describe a neural network. Finally, multiple neural networks will be applied to the automobile insurance data set to determine which neural network provides the best-fit model.

6.1 What Is a Neural Network?

Neural networks are a powerful analytic tool that seeks to mimic functions of the human brain. A key component is their ability to learn from experience. Within the brain, *neurons* are the component that enables cognition and intelligence. The brain is composed of a system (network) of neurons that work together to form one cohesive unit. Inputs arrive to each neuron through a connection called a dendrite. Dendrites transmit their information to the neuron by sending neurotransmitters across a synaptic gap. *These* neurotransmitters either excite or inhibit the receiving

R. V. McCarthy et al., *Applying Predictive Analytics*, https://doi.org/10.1007/978-3-030-83070-0_6

neuron. If they excite the receiving neuron, this is referred to as firing the neuron. If they inhibit the neuron, it does not become active. It is also important to note that the amount of the neurotransmitters that are transmitted across the synaptic gap determines the relative strength of each dendrite's connection to their corresponding neuron. This relative strength can be expressed as the *weight* of the input. The weights of the inputs are summed. If the sum of the weights exceeds an acceptable threshold level, then it causes the neuron to fire, sending a pulse to other neurons. This threshold level is referred to as a *bias*. The more active the synapse is, the stronger the connections, the weaker the synapse, and the more likely the neuron atrophies from lack of use.

Neural networks use this same concept to mirror the biological functions of the brain. A neural network consists of neurons with input, hidden layer(s), and output connections. Figure 6.1 presents a simple model of a neural network.

These connections are weighted and have a bias. They have the ability to be fired and thus utilized, depending upon the inputs. Those that are fired remain active, and those that are not remain inactive. Neural networks may be simple or highly complex. Like a brain, they have the ability to learn from experience (which in predictive analytics is historical data). Thus, a neural network can be a very useful predictive technique when historical data is representative of future results. Neural networks have the ability to process a vast amount of data. They can also be used on problems involving more than one target (dependent) variable. Neural networks adjust to new data, and thus they are adaptable. They also can deal with noisy or fuzzy data (i.e., inconsistent or incomplete data).

Neural networks learn from experience through categorization-based machine learning and are useful for pattern recognition-based problems. To accomplish this, they must be trained. Like the human brain, they *learn* from experience. Training takes place through the use of iterations of data being processed within the network. The more factors that are involved for the network to process, the more nodes (or units) that are needed (and usually the longer it will take to train the network). As is

Fig. 6.1 Neural network with one hidden layer

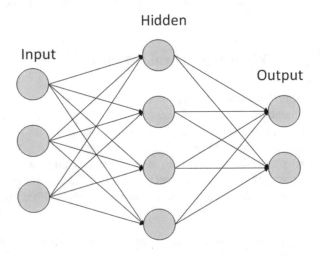

the case with decisions made by the human brain, some factors have more influence (weight) than others. Increasing the number of nodes and adding weights to those nodes increase the complexity of the network and add to the computational requirements (see Fig. 6.2). It can also increase the amount of time it takes for the network to train. Though training may take time, once a neural network is trained, it can be one of the fastest predictive modeling techniques. Therefore, they are frequently used today in applications such as voice response systems and fraud detection.

Neural networks have many applications in business today. They are used in sales forecasting, customer retention and identification of potential new customers, risk management, and industrial process control. Neural networks are used in credit scoring and credit card fraud applications. They are used widely in applications that make recommendations to customers on complimentary products. They have even begun to move into the healthcare market in applications such as personal heath monitoring. Neural networks work best in situations where prior data is both available and indicative of future events.

6.2 History of Neural Networks

The concept of an artificial neural network pre-dates our ability to automate them through specialized computer software. The concept was first put forth by McCulloch and Pitts (1943). They provided theorems to prove that networks of neurons could take any set of binary inputs and transform them into a set of binary outputs. They demonstrated that artificial neural networks are able to model input-output relationships to any degree of precision. McCulloch and Pitts (1943) also purported that the overall structure of the neural network does not change over time; however the interaction between the neurons changes. They based their assumptions on the following:

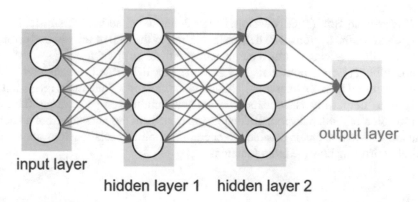

Fig. 6.2 Neural network with two hidden layer

- The activity of a neuron is an "all-or-none" process.
- A fixed number of synapses must be excited to excite a neuron. The synapses are independent of the previous position of the neuron and its activity.
- Synaptic delay is the significant delay.
- Neuron excitation can be prevented by the presence of inhibitory synapses.
- Neural network structures do not change over time (McCulloch and Pitts 1943).

A McCulloch-Pitts neuron is the foundational component of the McCulloch-Pitts network. Within the network, each input [x_1] is weighted [w_1] and summed. An overall bias [w_0] is added. This results in the neuron having either a positive or negative value which represents the networks' net result. This is expressed as a linear regression equation as follows:

$$\hat{y} = f\left(w_0 + \sum_{i=1}^{d} w_i x_i \right)$$

The weight associated with each input is the slope parameter, and the bias term is the y-intercept. Their output is determined by passing the input values through a binary step function. Their model formed the basis for the first computerized neural networks, but as is often the case, others then sought to improve upon it.

In 1949, Canadian neuropsychologist Donald Hebb put forth his theory of Hebbian learning in his book *The Organization of Behavior*. Hebb (1949) stated that:

> When an axon of cell A is near enough to excite cell B and repeatedly or persistently takes part in firing it, some growth process or metabolic change takes place in one or both cells such that A's efficiency, as one of the cells firing B, is increased.

This meant that the combination of neurons could be processed together and that the strength of the connection between two neurons could be adjusted. This results in the neurons' learning rate. If the rate of adjustment is set too low however, it may result in very lengthy training times. If, however, it is set to high, then it may result in divergence from the desired solution. As a result, Hebb's learning rate results in weights that are unbounded; this instability is considered a flaw that requires correction.

Widrow and Hoff (1960) developed their delta rule to address this instability in the weights of the neurons. The delta rule minimizes the sum of the squared error by replacing the receiving neuron's output value with an error value.

Rosenblatt (1958) implemented a variation of the delta rule in a computer model that he referred to as a *perceptron*. Though his initial model had input layers linked directly to output layers, he experimented with multilayer perceptrons (i.e., the introduction of hidden layers). Hidden layers transform the network from a simple linear model to one that is capable of generating nonlinear relationships, thus more closely mirroring how the brain functions.

6.3 Components of a Neural Network

The basis for neural network models is the inputs, hidden layer(s), and outputs. In SAS Enterprise Miner™, the outputs are the estimates or predictions for the target variables. Neural networks are capable of handling data sets that have multiple target variables. However, the type of data will determine the resulting output. A categorical (class) target will result in a probability, whereas an interval target will result in an expected value. As is the case with all analytic models, it is important to first consider the data and then transform variables that require adjustment. Neural networks have the advantage of acting as a complex set of nonlinear models that will transform the variables as they generate weights (model estimations).

The hidden layers perform mathematical calculations on the inputs they receive from the previous layer (either the input layer or a layer of hidden nodes). In SAS Enterprise Miner™, the formulas used by the hidden layers are known as the hidden layer combination function (i.e., this function is what determines how the inputs to the hidden layer nodes are combined). SAS Enterprise Miner™ also permits you to specify a target layer combination function to define how the inputs will be combined (see Table 6.1) to get to the output layer.

The hidden layer activation function is then processed to transform the combined values. The values that result from the generation of these activation and combination functions are the output of the node. The node output then serves as the input to the next node (either a hidden layer or a target/output) layer in the neural network model. For hidden layers, the default activation function depends upon the number of layers and the combination function that was used. As was the case with the activation function, SAS Enterprise Miner™ also includes a target combination function (see Table 6.2). The combination of the target activation and combination functions results in the final output. The functions that are specified should support

Table 6.1 SAS Enterprise Miner™ combination functions (support.sas.com)

Combination function	Definition
Add	Adds all of the incoming values without using any weights or biases
Linear	Is a linear combination of values and weights
EQ slopes	Identical to linear, except the same connection weights are used for each unit in the layer, although different units have different biases (mainly used for ordinal targets)
EQ radial	Is a radial basis function with equal heights and widths for all units in the layer
EH radial	Is a radial basis function with equal heights but unequal widths for all units in the layer
EW radial	Is a radial basis function with equal widths but unequal heights for all units in the layer
EV radial	Is a radial basis function with equal volumes for all units in the layer
X radial	Is a radial basis function with unequal heights and widths for all units in the layer

Table 6.2 SAS Enterprise Miner™ activation functions (support.sas.com)

Activation function
Identity
Linear
Exponential
Reciprocal
Square
Logistic
MLogistic
SoftMax
Gauss
Sine
Cosine
Elliott
TanH
Arctan

the target variable (i.e., since the target variable represents the desired answer/result for the business problem, it is important to align the level (data type) of the target to the combination function that is selected.

If the neural network has no hidden layers, then the identity function is the default. When hidden layers are present, the default hidden layer activation function is the TanH function. The logistic, hyperbolic tangent (TanH), Elliott, and Arc tangent (Arctan) functions are all sigmoid activation functions and perform relatively equivalently in most cases. Sigmoid functions can be helpful when dealing with data that is nonlinear because the output can be smoothed. They are particularly useful when generating probabilities as they return a result between 0 and 1. The TanH function can be especially useful when handling the impact of negative numbers, as it creates values between -1 and 1. The exponential function is useful for data distributions that are undefined at negative values, such as a Poisson distribution. The exponential function generates values that range from 0 to ∞. The SoftMax activation function works best with a multinomial distribution of the target. When the target distribution is normally distributed, the identity function is appropriate. Unlike the other functions, the identity function does not transform its argument so it can easily handle a distribution range from $-\infty$ to ∞. Identity is the default target activation function.

Neural networks also utilize an error function. The choice of the error function is driven by the level (data type) of the target as it impacts the parameter estimate that will result. The goal is to use an error function that will minimize this parameter estimate. Table 6.3 provides a summary of each of the error functions.

Table 6.3 Error functions

Normal	Commonly used with interval targets that are normally distributed. This can be used with any type of target variable to predict conditional means. This is also known as the least squares or mean squared error. This may be used for categorical targets that contain outliers
Cauchy	This can be used with any type of target variable to predict conditional modes
Logistic	This can be used with any type of target variable, but it is most commonly used with interval targets to mitigate the impact of outliers
Huber	This can be used for categorical targets to predict the mode rather than posterior probability. For an interval target, it is used for unbounded variables that may contain outliers
Biweight	This can be used with any type of target variable; however, it is most commonly used with interval targets that contain severe outliers. This will predict one mode of a multimodal distribution
Wave	This can be used with any type of target variable; however, like the biweight, it is most commonly used with interval targets that contain severe outliers. This will also predict one mode of a multimodal distribution
Gamma	This can be used only with interval target variables that have positive values. It is used when the standard deviation is proportional to the mean of the target variable. It is commonly used with target variables that have a skewed distribution
Poisson	This is used for interval targets that are non-negative and skewed, particularly counts of rare events where the conditional variance is proportional to the conditional mean
Bernoulli	This is only used with binary target variables
Entropy	This is used for interval target variables, whose value ranges between 0 and 1
MBernoulii	(Multiple Bernoulli) This can be used for categorical target variables
Multinomial	This is used when there are two or more non-negative interval target variables, where each case represents the number of trials that are equal to the sum of the target values. This is used with a SoftMax activation function to force the outputs to sum to 1
Mentropy	(Multiple Entropy, also referred to as the Kullback-Leibler divergence) This is used when there are two or more interval targets whose values are between 0 and 1. This is used with a SoftMax activation function to force the outputs to sum to 1

6.4   Neural Network Architectures

SAS Enterprise Miner™ supports several types of neural network architectures. The simplest is the generalized linear model (GLM). The generalized linear model creates a simple network with a default of three hidden nodes. It supports linear (default), additive (Add), and equal slopes combination functions. The additive function is rarely used and typically only performs better when then the inputs have been transformed. The equal slopes function is appropriate when the target variable is ordinal. Two limitations of the generalized linear model are that they can only solve problems that are linearly separable and it fails when there are too many inputs.

The most commonly used neural network model is the *multilayer perceptron* (MLP). Multilayer perceptron neural networks use historical data to produce a known output. The inputs are fed into the input layer where they are multiplied by the weights associated with their interconnections as they are passed to the first of one or more hidden layers. Within each hidden layer, they get summed and processed by a nonlinear function, where they are then passed to the next interconnected hidden node. Once they have been summed and processed through the hidden layers, they are then summed and processed one final time in the output layer resulting in the neural network output. Multilayer perceptrons rely on historical data to map the input to the output through the hidden layers. In this context, *the past must be indicative of the future*, which is to say that the historical data that is used to develop and train this network must be representative of the current data that will be processed through the network.

Historically, when there were more than two hidden layers, the number of hidden layers had a substantial impact on the training time needed by the neural network. Current neural network research referred to as *deep learning* (Hinton et al. 2006) has demonstrated that having more than two hidden layers may substantially improve the results of the neural network. The layers in this case are typically trained in pairs which help mitigate the training problem.

Multilayer perceptrons (as well as several other types of neural networks) are trained (i.e., learn) through *backpropagation*. Backpropagation is an algorithm. The input data is processed through the network, and the generated output is then compared to the desired output; the comparison results in the generation of an error. The error is then fed back into the neural network and used to adjust the weights for each node to continuously decrease the error with each iteration. The result is to move the generated output closer to the desired output. Each iteration is a training iteration. The training iteration that produces the generated result that is closest to the desired result becomes the initial version of the neural network model. This contains weights that most closely approximate the desired output.

It is possible to add a direct connection from the input layer to the output layer (bypassing the hidden layer(s)). This is referred to as a *skip-layer network* . In general, however, there is little value gained by doing this. The exception to this is when dealing with nonstationary data because the multilayer perceptron neural network is a stationary model.

The optimal number of hidden units (i.e., hidden node within a hidden layer) in a neural network is problem specific and more difficult to manage than the number of hidden layers. If a neural network has to many hidden units, it will model random variation (referred to as noise). This results in a neural network that fails to generalize. However, if there are too few hidden units, then the opposite occurs. The neural network will fail to capture the approximation of the desired result also resulting in a failure to generalize. Neural networks that fail to generalize lose their usefulness as they only work well with their training data. Principe et al. (2000) suggest the number of units in the first hidden layer should be about twice the number of input variables. The appropriate number of hidden units can then be determined empirically. Begin by measuring its performance with an appropriate metric. Increase the number of hidden units by one, to determine the impact on the network's fit until the performance of the network begins to drop. The optimal number of hidden units will be the number associated with the best-fit performance.

One important consideration when using neural networks is that the data must be encoded by the network. When dealing with categorical variables therefore, it is important to consider which categories are important. It may be necessary to consolidate or eliminate some categories from consideration, particularly if they are not essential to the issue being analyzed. The values must have a numeric equivalent, and the network only knows those values, not the underlying meaning. Typically values between 0 and 1 will be assigned. For example, if we have a category that represents gender and we have three values (F for female. M for male, and U for unknown), the network may represent those as 0, 1, and 0.5. In the data preparation stage, it is important to consider the number of categories in each categorical variable. The output nodes are also computed as a numeric value between 0 and 1. For a binary target, this of course makes it easy to interpret the output. For interval targets, it may be necessary to evaluate this result. When the output is a probability, this is directly interpretable based upon the calculated value of the output.

6.5   Training a Neural Network

How does a neural network learn? Neural networks require a training data set. The training data needs to be sufficiently large enough for the network to be able to adjust the calculated values of each of the nodes. In general, bigger is better (although a network can be overfit). In today's world of big data, it has become less of a consideration to be able to acquire sample sizes large enough to train neural networks. Each observation (i.e., row or record) from the training data set is processed through the network, and an output value (i.e., target) is calculated. The output values that were calculated are then compared to the original output value from the training record, and the difference (actual value- calculated value) is the error. Using backward propagation rules, the network can then take the errors back through the network for the purpose of adjusting the weights to minimize the error. In this way the network is considered to be able to learn from itself (also referred to as supervised learning). Multiple iterations of the back propagation are needed to train the network. How many iterations? It depends upon the data, but in all cases at some point, the network will hit a point where the improvement diminishes, and thus the network is considered to be trained.

Before you train a neural network, it is best to prepare the data set. Use only the important variables; when the data set is too large, it may take a very long time to train. For example, if the data contains thousands of variables, it may be best to pick the 20–50 most important variables. Partitioning should be used. The training data set is used for the neural network to be able to learn the network weights. The validation data set, which is typically how the best-fit model is decided, is used to determine which of the various network architectures will be used in the model. If a test data set is created, it is used to create a final unbiased estimate to improve the generalization of the neural network.

6.6   Radial Basis Function Neural Networks

Radial basis function (RBF) neural networks may offer a useful alternative to either the generalized linear model or multilayer perceptron model. They were first proposed by Broomhead and Lowe (1988) and differ from the multilayer perceptron because they use radial basis functions (Gaussians, i.e., exponential or SoftMax) instead of sigmoid functions as their hidden layer activation function. The combination function acts as a distance function within this network. It is typically the squared Euclidean distance between the weight vector and the input vector. They use linear combination functions in the output layer. The closer the inputs are to the outputs, the stronger the radial basis function will be. In two dimensions, all inputs that are the same distance from the output will form a circle; in three dimensions they will form a sphere. There are no weights associated with the connections between the inputs and the hidden units. The shorter the distance between the inputs and the outputs, the more reliable the network will be.

Radial basis function networks rapidly train and are known for their ability to generalize. This can be an important reason to utilize them, particularly when dealing with high-volume data sets. As a result, they can be useful for problems that involve function approximation, time series prediction, and classification. They work with any number of inputs and typically have one hidden layer with any number of units. There are two types of radial basis function networks:

- *Ordinary radial basis functions* (ORBF): These utilize an exponential activation function to generate a Gaussian within the hidden layer weights which serves to provide a "bump" to the input.
- *Normalized radial basis functions* (NRBF): These utilize a SoftMax activation function which is the product of the cooperative of all participating hidden units and sums to 1.

Normalized radial basis function neural networks usually provide a better-fit neural network than ordinary radial basis function networks; however, as is always the case, this should be tested.

There are five combination functions that can be used with radial basis function networks. They are:

1. XRadial—This is the most memory intensive.
2. EQRadial—This is the least memory intensive.
3. EWRadial
4. EHRadial
5. EVRadial—This is the least reliable and is rarely used.

6.7   Creating a Neural Network Sing SAS Enterprise Miner™

Let's examine the use of neural networks as a predictive analytics tool. We will apply several neural networks to the automobile insurance claim data set and examine the results to see which neural network provides the best predictive capability.

The first neural network that will be utilized will be the generalized linear model (Fig. 6.3). The neural network nodes are located within the Model table of the SEMMA model. The node has been renamed from its default. The neural network node will utilize the results of the data preparation and transformation. The transformed data provides a more reliable input for the neural networks.

Now let's examine the properties of this neural network (Fig. 6.4).

Note the number of hidden units default to 3 and cannot be changed. For this example, the default settings, which include a default target activation, combination, and error function, will be used. Consistent with our prior analysis using decision trees and regression, we will set the model selection criterion to average error (Fig. 6.5).

Figure 6.6 provides the fit statistics. Notice the Fit Statistics window provides several selection statistics including our selected statistic: average error.

The iteration plot (Fig. 6.7) within the *Results* (from SAS Enterprise Miner™) provides the number of iterations required to train the neural network using this data. In this case it was ten iterations. This model stabilizes very quickly, the average error shows no improvement after the first iteration, and other measures require more iterations but stabilize by the tenth iteration.

The results from the neural network also provide a graph of the cumulative lift of both the train and validation data sets (Fig. 6.8). The cumulative lift chart provides a visual representation of how well the neural network performs versus random guessing.

Our simple neural network model produced a reasonable result. The average squared error is within a relevant range with our other predictive techniques. However, what happens when the target activation and combination functions are set to linear (Fig. 6.9)?

Fig. 6.3 Neural
network node

Fig. 6.4 Generalized linear model properties

In this case the resulting average squared error is now 0.061225 (Fig. 6.10), which is higher than using the default target activation and combination functions. This network produced a worse result; therefore it will not be considered any further in our analysis.

Since the multilayer perceptron architecture is so commonly used, let's compare the results of an MLP neural network to the GLM neural network. Again, two examples to illustrate the movement of the fit statistic and determine which of these will provide the best fit will be used. In the first example (Fig. 6.11), the default properties will be used for the multilayer perceptron neural network.

Fig. 6.5 Generalized
linear model selection
criterion

Fig. 6.6 Fit statistics for the generalized linear model neural network

Target	Target Label	Fit Statistics	Statistics Label	Train	Validation
Fraudulent_Claim	Fraudulent_Claim	_DFT_	Total Degrees of Freedom	2997	
Fraudulent_Claim	Fraudulent_Claim	_DFE_	Degrees of Freedom for Error	2971	
Fraudulent_Claim	Fraudulent_Claim	_DFM_	Model Degrees of Freedom	26	
Fraudulent_Claim	Fraudulent_Claim	_NW_	Number of Estimated Weights	26	
Fraudulent_Claim	Fraudulent_Claim	_AIC_	Akaike's Information Criterion	1231.085	
Fraudulent_Claim	Fraudulent_Claim	_SBC_	Schwarz's Bayesian Criterion	1387.225	
Fraudulent_Claim	Fraudulent_Claim	_ASE_	Average Squared Error	0.054129	0.055389
Fraudulent_Claim	Fraudulent_Claim	_MAX_	Maximum Absolute Error	0.981207	1
Fraudulent_Claim	Fraudulent_Claim	_DIV_	Divisor for ASE	5994	4004
Fraudulent_Claim	Fraudulent_Claim	_NOBS_	Sum of Frequencies	2997	2002
Fraudulent_Claim	Fraudulent_Claim	_RASE_	Root Average Squared Error	0.232656	0.23535
Fraudulent_Claim	Fraudulent_Claim	_SSE_	Sum of Squared Errors	324.4485	221.7794
Fraudulent_Claim	Fraudulent_Claim	_SUMW_	Sum of Case Weights Times Freq	5994	4004
Fraudulent_Claim	Fraudulent_Claim	_FPE_	Final Prediction Error	0.055076	
Fraudulent_Claim	Fraudulent_Claim	_MSE_	Mean Squared Error	0.054603	0.055389
Fraudulent_Claim	Fraudulent_Claim	_RFPE_	Root Final Prediction Error	0.234683	
Fraudulent_Claim	Fraudulent_Claim	_RMSE_	Root Mean Squared Error	0.233672	0.23535
Fraudulent_Claim	Fraudulent_Claim	_AVERR_	Average Error Function	0.196711	0.228071
Fraudulent_Claim	Fraudulent_Claim	_ERR_	Error Function	1179.085	913.1963
Fraudulent_Claim	Fraudulent_Claim	_MISC_	Misclassification Rate	0.061061	0.061938
Fraudulent_Claim	Fraudulent_Claim	_WRONG_	Number of Wrong Classifications	183	124

The default number of hidden units is 3, and the default direct connection (input layer to output layer) is set to No (i.e., this is not a skip-layer architecture). This is a rather simple multilayer perceptron network. The results of this network show an average squared error of 0.055166 (Fig. 6.12).

After the tenth training iteration, the network stabilizes (Fig. 6.13), and further training (up to the default of 50) does not result in significant gain.

Note the difference in the cumulative lift (Fig. 6.14) compared to the generalized linear model architecture (Fig. 6.8). The validation cumulative lift for the multilayer perceptron model is higher than it was for the generalized linear model. This does not mean that the multilayer perceptron model will always be the superior

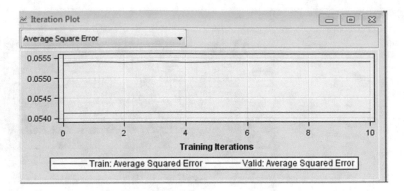

Fig. 6.7 Iteration plot for the generalized linear model neural network

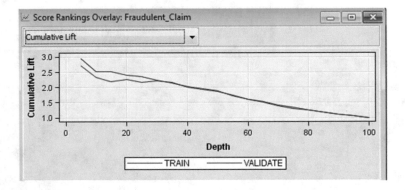

Fig. 6.8 Cumulative lift of the generalized linear model neural network

architecture for every business problem; the results will vary based upon the data used to train the networks.

A final neural network will be examined. In this case the neural network will continue to use the multilayer perceptron architecture; however this time the number of hidden units will increase from its default (3) to 6 (Fig. 6.15). This will cause an increase in the complexity of the network.

To account for the increase number of relationships between the nodes, the number of training iterations will be increased from its default (50) to 150 (Fig. 6.16).

This change in the complexity of the neural network has an impact on the results. In this case, the average squared error is now 0.056142 (Fig. 6.17). This is a slight increase from the three hidden unit networks, indicating that this network performs slightly worse and therefore it will not be considered any further.

Note that in this case the cumulative lift of the validation data set (Fig. 6.18) is significantly less than the cumulative lift of the model with only three hidden units. Adding more hidden units does not always mean it will improve the network. If this were the case, then we would always start out with large, complex models (And work our way up!).

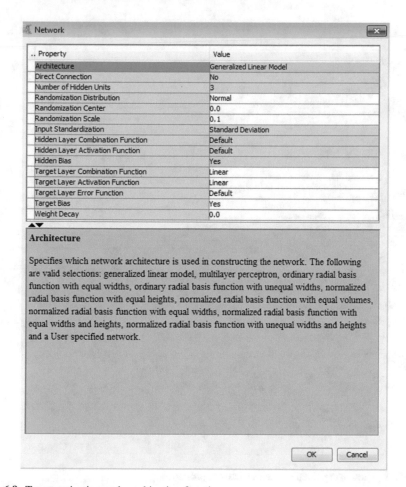

Fig. 6.9 Target activation and combination functions

Target	Target Label	Fit Statistics	Statistics Label ▲	Train	Validation
Fraudulent_Claim	Fraudulent_Claim	_AIC_	Akaike's Information Criterion	965.5863	
Fraudulent_Claim	Fraudulent_Claim	_AVERR_	Average Error Function	0.143741	1852.957
Fraudulent_Claim	Fraudulent_Claim	_ASE_	Average Squared Error	0.060458	0.061225
Fraudulent_Claim	Fraudulent_Claim	_DFE_	Degrees of Freedom for Error	2945	
Fraudulent_Claim	Fraudulent_Claim	_DIV_	Divisor for ASE	5994	4004
Fraudulent_Claim	Fraudulent_Claim	_ERR_	Error Function	861.5863	7419240
Fraudulent_Claim	Fraudulent_Claim	_FPE_	Final Prediction Error	0.062593	
Fraudulent_Claim	Fraudulent_Claim	_MAX_	Maximum Absolute Error	0.995879	0.993112
Fraudulent_Claim	Fraudulent_Claim	_MSE_	Mean Squared Error	0.061526	0.061225
Fraudulent_Claim	Fraudulent_Claim	_MISC_	Misclassification Rate	0.061061	0.061938
Fraudulent_Claim	Fraudulent_Claim	_DFM_	Model Degrees of Freedom	52	
Fraudulent_Claim	Fraudulent_Claim	_NW_	Number of Estimated Weights	52	
Fraudulent_Claim	Fraudulent_Claim	_WRONG_	Number of Wrong Classifications	183	124
Fraudulent_Claim	Fraudulent_Claim	_RASE_	Root Average Squared Error	0.245882	0.247437
Fraudulent_Claim	Fraudulent_Claim	_RFPE_	Root Final Prediction Error	0.250186	
Fraudulent_Claim	Fraudulent_Claim	_RMSE_	Root Mean Squared Error	0.248043	0.247437
Fraudulent_Claim	Fraudulent_Claim	_SBC_	Schwarz's Bayesian Criterion	1277.865	
Fraudulent_Claim	Fraudulent_Claim	_SUMW_	Sum of Case Weights Times Freq	5994	4004
Fraudulent_Claim	Fraudulent_Claim	_NOBS_	Sum of Frequencies	2997	2002
Fraudulent_Claim	Fraudulent_Claim	_SSE_	Sum of Squared Errors	362.3856	245.1453
Fraudulent_Claim	Fraudulent_Claim	_DFT_	Total Degrees of Freedom	2997	

Fig. 6.10 Results with target activation and combinations functions set to linear

Fig. 6.11 Multilayer perceptron neural network properties

Target	Target Label	Fit Statistics	Statistics Label	Train	Validation
Fraudulent_Claim	Fraudulent_Claim	_DFT_	Total Degrees of Freedom	2997	
Fraudulent_Claim	Fraudulent_Claim	_DFE_	Degrees of Freedom for Error	2915	
Fraudulent_Claim	Fraudulent_Claim	_DFM_	Model Degrees of Freedom	82	
Fraudulent_Claim	Fraudulent_Claim	_NW_	Number of Estimated Weights	82	
Fraudulent_Claim	Fraudulent_Claim	_AIC_	Akaike's Information Criterion	1236.665	
Fraudulent_Claim	Fraudulent_Claim	_SBC_	Schwarz's Bayesian Criterion	1729.105	
Fraudulent_Claim	Fraudulent_Claim	_ASE_	Average Squared Error	0.051626	0.055166
Fraudulent_Claim	Fraudulent_Claim	_MAX_	Maximum Absolute Error	0.994303	0.999904
Fraudulent_Claim	Fraudulent_Claim	_DIV_	Divisor for ASE	5994	4004
Fraudulent_Claim	Fraudulent_Claim	_NOBS_	Sum of Frequencies	2997	2002
Fraudulent_Claim	Fraudulent_Claim	_RASE_	Root Average Squared Error	0.227215	0.234874
Fraudulent_Claim	Fraudulent_Claim	_SSE_	Sum of Squared Errors	309.4491	220.8846
Fraudulent_Claim	Fraudulent_Claim	_SUMW_	Sum of Case Weights Times Freq	5994	4004
Fraudulent_Claim	Fraudulent_Claim	_FPE_	Final Prediction Error	0.054531	
Fraudulent_Claim	Fraudulent_Claim	_MSE_	Mean Squared Error	0.053079	0.055166
Fraudulent_Claim	Fraudulent_Claim	_RFPE_	Root Final Prediction Error	0.233519	
Fraudulent_Claim	Fraudulent_Claim	_RMSE_	Root Mean Squared Error	0.230388	0.234874
Fraudulent_Claim	Fraudulent_Claim	_AVERR_	Average Error Function	0.178956	0.202123
Fraudulent_Claim	Fraudulent_Claim	_ERR_	Error Function	1072.665	809.3022
Fraudulent_Claim	Fraudulent_Claim	_MISC_	Misclassification Rate	0.061061	0.061938
Fraudulent_Claim	Fraudulent_Claim	_WRONG_	Number of Wrong Classifications	183	124

Fig. 6.12 Multilayer perceptron results

Fig. 6.13 Multilayer perceptron iteration plot

Fig. 6.14 Multilayer perceptron cumulative lift

From the training iteration plot (Fig. 6.19), note that this model trains very quickly. Adding more training iterations was inconsequential to the training of this network. In fact, it merely increased the run time to train the network with no additional benefit. Rather than increase the number of training iterations, the number of training iterations could have been decreased and achieved the same result. This may be an important consideration when training a network with a very large data set.

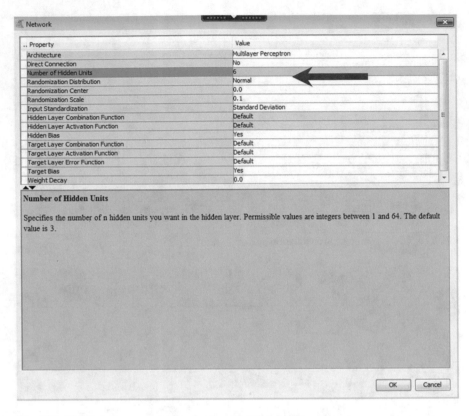

Fig. 6.15 Multilayer perceptron neural network with six hidden units

6.8   Using SAS Enterprise Miner™ to Automatically Generate a Neural Network

The AutoNeural network node (Fig. 6.20) will automatically configure a neural network. This can be particularly useful for someone new to predictive analytics who is just getting started with using neural networks.

There are a limited number of properties that can be adjusted when using the AutoNeural node. The default termination for this neural network node causes the network to stop training once it has hit overfitting. This can be modified to either a time limit or training error. The number of hidden units can be adjusted from its default (two). Hidden units are added to the network one at a time. In addition, the maximum iterations and time limit can also be adjusted. The target error function can be changed; however the combination and activation functions are automatically set by this node. When the Train action property is set to Search, the AutoNeural node will search for the best optimized model.

There are four types of architectures supported by the AutoNeural node:

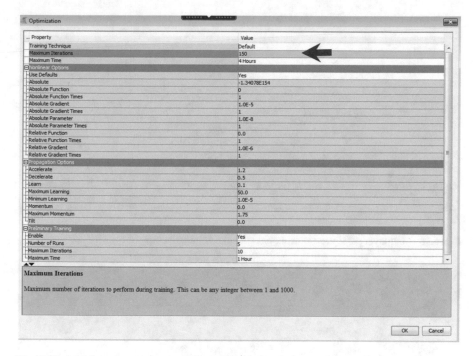

Fig. 6.16 Multilayer perceptron number of iterations

Target	Target Label	Fit Statistics	Statistics Label	Train	Validation
Fraudulent_Claim	Fraudulent_Claim	_DFT_	Total Degrees of Freedom	2997	
Fraudulent_Claim	Fraudulent_Claim	_DFE_	Degrees of Freedom for Error	2834	
Fraudulent_Claim	Fraudulent_Claim	_DFM_	Model Degrees of Freedom	163	
Fraudulent_Claim	Fraudulent_Claim	_NW_	Number of Estimated Weights	163	
Fraudulent_Claim	Fraudulent_Claim	_AIC_	Akaike's Information Criterion	1439.52	
Fraudulent_Claim	Fraudulent_Claim	_SBC_	Schwarz's Bayesian Criterion	2418.394	
Fraudulent_Claim	Fraudulent_Claim	_ASE_	Average Squared Error	0.052491	0.056142
Fraudulent_Claim	Fraudulent_Claim	_MAX_	Maximum Absolute Error	0.999206	0.999002
Fraudulent_Claim	Fraudulent_Claim	_DIV_	Divisor for ASE	5994	4004
Fraudulent_Claim	Fraudulent_Claim	_NOBS_	Sum of Frequencies	2997	2002
Fraudulent_Claim	Fraudulent_Claim	_RASE_	Root Average Squared Error	0.22911	0.236942
Fraudulent_Claim	Fraudulent_Claim	_SSE_	Sum of Squared Errors	314.6334	224.7906
Fraudulent_Claim	Fraudulent_Claim	_SUMW_	Sum of Case Weights Times Freq	5994	4004
Fraudulent_Claim	Fraudulent_Claim	_FPE_	Final Prediction Error	0.05853	
Fraudulent_Claim	Fraudulent_Claim	_MSE_	Mean Squared Error	0.05551	0.056142
Fraudulent_Claim	Fraudulent_Claim	_RFPE_	Root Final Prediction Error	0.241929	
Fraudulent_Claim	Fraudulent_Claim	_RMSE_	Root Mean Squared Error	0.235607	0.236942
Fraudulent_Claim	Fraudulent_Claim	_AVERR_	Average Error Function	0.185772	0.205201
Fraudulent_Claim	Fraudulent_Claim	_ERR_	Error Function	1113.52	821.6242
Fraudulent_Claim	Fraudulent_Claim	_MISC_	Misclassification Rate	0.061061	0.063936
Fraudulent_Claim	Fraudulent_Claim	_WRONG_	Number of Wrong Classifications	183	128

Fig. 6.17 Multilayer perceptron with six hidden units results

- *Single layer*—The specified number of hidden units is added to the existing hidden layer in parallel. If the specified units are set to a number greater than 1, then the units are added in groups.
- *Block layer*—Hidden units/layers are added as additional layers within the network. All of the layers within the network have the same number of units. This number is specified in the Number of Hidden Units property. Because the number of weights grows as you add each additional layer, this architecture is not as commonly used.

Fig. 6.18 Multilayer perceptron with six hidden units cumulative lift

Fig. 6.19 Multilayer perceptron with six hidden units iteration plot

Fig. 6.20 AutoNeural
network node

- *Funnel layer*—Multiple hidden layers are created, and when each new layer is created, a single unit is added to the prior layer forming a funnel pattern.
- *Cascade layer*—Each hidden unit is connected to all preceding hidden and input units. As a result, the network is built one unit as a time.

The DMNeural node (Fig. 6.21) may be useful when your target variable is binary or interval. It uses an additive nonlinear model that was developed to overcome the problem nonlinear estimation, computing time, and finding global optimal solutions.

Fig. 6.21 DMNeural
network node

Nonlinear estimation problems can occur when the neural network is unable to determine at what point the networks estimate is close enough to the optimal solution. The DMNeural node trains segments of data at a time in order to reduce the computing time needed to train the neural network. Some neural networks may require an extensive amount of training time due to the number of iterations necessary to complete the training process. Many neural network solutions are very sensitive to the specific data they were trained with and thus find an optimal local solution that may not be as generalizable. The DMNeural node seeks to find a more global generalizable solution. The DMNeural node uses each of eight separate activation functions to determine which one works best. The combination function will default to IDENT for binary target and LOGIST for an interval target. These are the only types of target variables supported by this function. In addition, the DMNeural node requires at least two input variables.

One of the criticisms of neural networks is that they take too long to execute. This may be especially true in today's world of big data where instead of small samples, we now have access to an entire population of data. One of the fundamental questions that predictive analytics seeks to answer is "what insights are contained within the data?" What is the point of terabyte or petabyte storage, if you cannot use the data that is within these massive data stores? The HPNeural node (Fig. 6.22) is one of several high-performance nodes that is supported by SAS Enterprise Miner™ for the purpose of processing big data.

The HPNeural node provides the ability to create a neural network capable of processing a large amount of data to be utilized for optimal decision-making. It supports high-volume processing by minimizing the amount of data movement. Newer database technologies take advantage of features such as in-line memory to reduce the amount of data that is read/written to storage devices and support analytic processing that runs in parallel.

Most properties in the HPNeural node are set automatically, including the activation and error functions and the standardizing of input and target variables. The input(s) and target(s) may be interval, binary, or nominal. If there is a missing value in any of the targets, that row of data will be ignored. The Train properties (Fig. 6.23) enable you to configure the neural network that will be generated.

The following properties are available to be modified:

1. Variables—Permits you to set which variables to Use/Don't Use within this network.
2. Use Inverse Priors—The default is No. Set this to Yes when you have a nominal target that has rare events. It will calculate a weight that is the inverse fraction of the number of times the target occurs in the data set.

Fig. 6.22 HPNeural node

Fig. 6.23 Train properties
of the HPNeural node

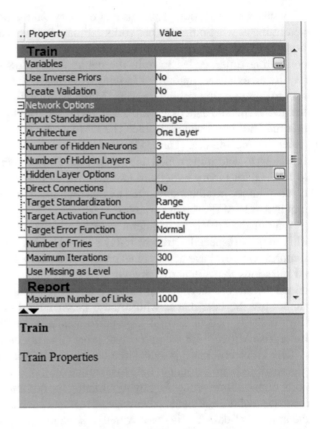

3. Create Validation—Specifies whether a validation data set will be created from the training data set. If you specified a validation data set using an HPDataPartition node, this will be ignored. Otherwise, every fourth record will be used in the validation data set that is created. Training will continue until there can be no further improvement in the training error or the number of iterations is reached.
4. Input Standardization—Identifies the method used to standardize the input variables (values are Range, Z Score, None).
5. Architecture—Permits you to define one of six architectures.

   (a) User Defined—When specified the Number of Hidden Layers, Hidden Layer Options, and Direct Connections can be modified.
   (b) Logistic—No hidden layers are created; logistic regression is used.
   (c) One Layer—A single hidden layer with no direct connections.
   (d) One Layer with Direct—A single hidden layer with direct connections.

(e) Two Layers—Two hidden layers with no direct connections.
(f) Two Layers with Direct—Two hidden layers with direct connections.

6. Number of Hidden Neurons—Specifies the number of hidden units in the net-work. The default is 3; if there are two hidden layers, then the number of units is split evenly between the two layers. If the number of hidden units is odd, then the first layer contains the extra hidden unit.
7. Number of Hidden Layers—Identifies the number of hidden layers in the net-work. The maximum is 10.
8. Hidden Layer Options—Permits you to specify either (or both) the number of hidden units per layer and the activation function.
9. Direct Connection—This applies only to the user-defined architecture. When set to Yes, it creates a direct link from the input to the output layer.
10. Target Standardization—Identifies the method used to standardize interval tar-get variables; the values are Range, Z Score, or None.
11. Target Activation Function—Identifies the target activation function for an interval target. The values are Identity, Exponential, Sine, and Tanh. Identity is required if the target standardization is set to None or Z Score.
12. Target Error Function—Identifies the target error function used for interval tar-get variables. The values are Normal, Poisson, or Gamma, If either Poisson or Gamma are utilized, then the target activation function must be set to Exponential. If the target error function is set to Normal, then the target activa-tion function must be either Tanh, Sine, or Identity.
13. Number of Tries—Specifies the number of unique attempts that will be tried to train the network utilizing a new set of weights for each attempt. The default is 2.
14. Maximum Iterations—Specifies the maximum number of iterations within each unique attempt that will be allowed.

The HPNeural node produces a Link Graph plot which is not available with the neural network plot. Figure 6.24 shows a sample Link Graph plot. The Link Graph

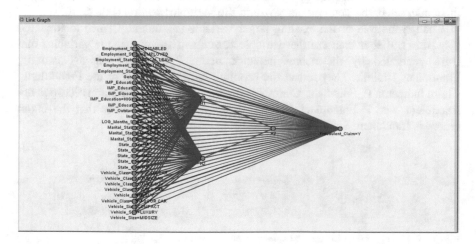

Fig. 6.24 Sample Link Graph plot

plot is a graphical representation of the neural network model. The target variable is always displayed on the right side, and all of the inputs are displayed on the left side of the diagram, with the hidden units reflected in between.

The HPNeural node provides a rapid method for developing powerful neural network models that are capable of handling big data data sets. They provide a means to enable the use of an entire data set instead of having to sample or segment the data.

6.9   Explaining a Neural Network

Neural networks can be a complex and powerful tool, and as a result, one of their criticisms is that they are difficult to understand. They are often thought of as a black box, or their output was produced by "the system." In many business problems however, it is not enough to merely produce an answer; the management of an organization may also want to know what were the underlying factors (i.e., what were the significant input variables) that led to the result.

Using SAS Enterprise Miner™, a neural network can be explained using a decision tree. This provides a reasonable explanation of the most important factors that were utilized by the neural network.

Let's see this in action by examining the multilayer perceptron neural network that contained three hidden units. This node was chosen because it was the best-performing model that was created using the claim fraud example.

To examine this neural network with a decision tree, two nodes will be utilized (see Fig. 6.25):

1. Metadata node
2. Decision Tree node

The Metadata node permits changes to the metadata for a SAS data set. Within this node, select the Train property within the Variables section (Fig. 6.26).

Change the role of the existing target variable (Fraudlent_Claim) to Rejected (Fig. 6.27). Rather than use that variable for consideration, use the variables that were generated by the neural network node, P_Fraudlent_ClaimN and P_Fraudlent_ClaimY. They reflect the probability of the values of the Fraudlent_Claim indicator. P_Fraudlent_ClaimN is the probability that the claim record is not fraudulent, and P_Fraudlent_ClaimY represents the probability that the claim record is fraudulent. Set these both to Target.

Fig. 6.25 Model for decision tree to explain a neural network

Fig. 6.26 Properties of the
Metadata node

Fig. 6.27 Metadata

Fig. 6.28 Subtree assessment plot

Fig. 6.29 Decision Tree
properties

.. Property	Value
┊Number of Surrogate Rules	0
┴Split Size	.
⊟Split Search	
┊Use Decisions	No
┊Use Priors	No
┊Exhaustive	5000
┴Node Sample	20000
⊟Subtree	
┊Method	N
┊Number of Leaves	6
┊Assessment Measure	Decision
┴Assessment Fraction	0.25
⊟Cross Validation	
┊Perform Cross Validation	No

The Decision Tree node can now be used to produce an initial result. Examine the subtree plot for the average squared error. The subtree plot produces 42 leaves (Fig. 6.28). However, after the sixth leaf, the remaining leaves do not have a significant impact on the output of the neural network.

Using this now as a guide, a decision tree can be produced that is easy to explain and closely approximates the significant factors for this neural network. To do so, change the Method property of the Decision Tree node from assessment to N and the number of leaves to 6 (Fig. 6.29).

Fig. 6.30 Decision tree results

This produces a decision tree that shows that the input variables that had the most significant impact on the neural network. The input variables are Gender, Employment_Status, Vehicle_Class, and Marital_Status (Fig. 6.30).

Recall that the color of the leaf indicates the degree to which that variable had an impact on the overall decision. Could the decision tree that included all 42 leaves have been used? Yes, but it would require that more branches and leaves are evaluated to get to an understanding of which variables had a significant impact. This provides a quick method to demystify neural network models and explain their results. You may also find it useful to identify which variables to remove from further analysis because they are not significant enough to support the cost of their inclusion.

6.10 Chapter Summary

Neural networks are considered to be one of the *Big 3* predictive analytics tools. Although the idea for their use began in the 1940s with McCulloch and Pitts, it has been much more recent that neural networks have exploded onto the analytics landscape. The factors that have supported this massive expansion of use have been big data and improvements in processing speed as well as in-line memory. Large volumes of data can be used to train a neural network for widespread use in voice response systems, for pattern recognition systems, and for predictive analytics.

A neural network consists of inputs, hidden layers, and outputs that seek to mimic the functions of the human brain. The inputs are the independent variables and may consist of interval, ordinal, binary, or categorical data types. Neural networks may have one or more target variables also consisting of these data types. In between the inputs and the outputs are more hidden layers that may contain a variety of different hidden units. Many neural networks use back propagation to derive a result. Back propagation consists of at least two passes through a network. The first pass through the network processes the inputs to create an initial output. Then the difference between the actual output and the desired output is used to generate an error signal that is propagated back through the network to teach it to come closer to the desired result. In this way the neural network learns from its data.

Three neural network architectures that are all supported by SAS Enterprise Miner™ were discussed. These included the generalized linear model, radial basis functions, and the most commonly used architecture, the multilayer perceptron. The generalized linear model creates a simple neural network with a default of three hidden units. Multilayer perceptrons rely on historical data to map the inputs through the hidden layers to the outputs. In today's world of big data, neural networks enable organizations to take advantage of the volumes of data available regarding their customers and operations to uncover aspects of their business that are otherwise unknown to them. Radial basis functions rapidly train and are known for their ability to generalize. They differ from the multilayer perceptron in their use of a Gaussian function instead of sigmoid functions as their activation function.

In addition, SAS Enterprise Miner™ provides support for neural networks that can be created with relatively few considerations to the structure and make-up of the architecture. These include the AutoNeural and DMNeural nodes. The AutoNeural node supports four simple architectures which are single layer, block layer, funnel layer, and cascade layer. The DMNeural node may be used when the target variable is binary or interval.

The HPNeural node is a high-performance neural network designed to support massive data stores by minimizing the amount of data movement and taking advantage of parallel processing and in-line memory.

One of the criticisms of neural networks is that they are difficult to explain. To demystify that criticism, we utilized a decision tree to explain the input variables that had the most significant impact on a neural network.

Discussion Questions

1. What is meant by the term *deep learning*?
2. There are many applications of neural networks used in businesses today. Discuss one example where an organization is using neural networks. What is the problem or issue they are addressing? How has the neural network helped them? What is the architecture of the network?
3. Discuss how a neural network can be optimized to improve its performance.
4. Compare the use of neural networks for predictive analytics with linear and logistic regression. What are the advantages and disadvantages of each technique?
5. SAS Enterprise Miner™ supports several different neural network architectures and permits a wide variety of options, based upon the setting of properties for these networks. Describe three properties that are part of the neural network node. What are their values? When is it appropriate to use them?
6. Explain what is meant by *back propagation*.

References

Broomhead D, Lowe D (1988) Multi-variable functional interpolation and adaptive networks. Complex Syst 2:327–355

Hebb D (1949) The organization of behavior: a neuropsychological theory. Wiley and Sons, New York

Hinton G, Osindero S, Teh Y (2006) A fast learning algorithm for deep belief nets. Neural Comput 18:1527–1554

McCulloch W, Pitts W (1943) A logical calculus of the ideas immanent in nervous activity. Bull Math Biophys 5:115–133

Principe J, Euliano N, Lefebvre W (2000) Neural and adaptive systems. Wiley and Sons, New York

Rosenblatt F (1958) The perceptron: a probabilistic model for information storage and organization in the brain. Psychol Rev 65:386–408

support.sas.com/documentation/onlinedoc/miner/em43/neural.pdf

Widrow B, Hoff M (1960) Adaptive switching circuits. technical report

Chapter 7
Model Comparisons and Scoring

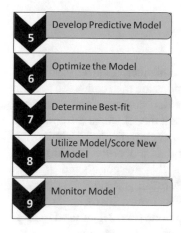

Chapter Learning Objectives
1. Create and utilize gradient boosting models.
2. Create and utilize ensemble models.
3. Evaluate multiple predictive models based upon fit statistics.
4. Utilize a best-fit predictive model to apply to current business situations.

7.1 Beyond the Big 3

Up to this point, we have focused on what is considered the Big 3 in predictive analytics: regression, decision trees, and neural networks. However, these are not the only methods available to us. Other methods have been developed, and their use has begun to become more widespread. The predictive analytics landscape has had a significant growth as we see more opportunities to apply these techniques in new and interesting applications. Business practices of yesteryear stressed giving the customer what they want. Business practices of tomorrow will focus on giving the customer what they need but aren't aware that they need it yet. There is what often seems like an infinite amount of data available, and we are no longer constrained by computer processors that were incapable of performing the calculations needed to evaluate a model in a timely manner. New predictive techniques will continue to be developed and utilized. Chapter 7 begins by examining some of the newer

© The Author(s), under exclusive license to Springer Nature
Switzerland AG 2022
R. V. McCarthy et al., *Applying Predictive Analytics*,
https://doi.org/10.1007/978-3-030-83070-0_7

predictive analytics techniques. It then covers a more in-depth discussion for evaluating different predictive models by evaluating fit statistics. The chapter then concludes with a review of scoring, the process used to apply a predictive model to new data.

7.2   Gradient Boosting

A boosting algorithm is one that is built in a stepwise manner. It makes predictions regarding a variety of different factors. These predictions are then combined to form a single, overall prediction. Gradient boosting uses decision trees and regression algorithms against a large amount of data to produce a model where the combined techniques may produce results that are superior to each individual technique. It typically handles outliers and missing values better than a decision tree or regression analysis would. There are multiple algorithms that are considered gradient boosting. One of the best-known algorithms is XGBoost. It is designed for parallel tree construction to increase the speed of execution (Browniee 2016).

In SAS Enterprise Miner™, the gradient boosting node (Fig. 7.1) is used to resample the analysis data set several times for the purpose of producing a weighted average of the resampled data set.

A series of decision trees are used to form a single predictive model. Multiple trees are used so that each successive tree can attempt to correct inaccuracies in the prior trees by fitting a residual (gradients of the error function) of the prediction from the earlier trees. The residual is defined as a derivative of the loss function (i.e., this shows the rate of change at a given point). Since each successive decision tree iteration is built upon the previous version of the decision tree, this is a stochastic model. Gradient boosting works with interval, nominal, and binary targets. If the target is an interval variable, the Huber M-Regression property (Fig. 7.2) can be set to change the loss function from the square error loss function. The Huber M-Regression loss function is less sensitive to outliers.

When the Huber M-Regression property is set to No, the square error loss function will be used. For squared error loss with an interval target, the residual is simply the target value minus the predicted value. Whereas one of the criticisms of neural networks is that they are difficult to explain, the gradient boosting provides several distinctive results. Using our automobile insurance claim data set, let's examine the

Fig. 7.1 Gradient
boosting node

Fig. 7.2 Gradient boosting
properties

.. Property	Value
General	
Node ID	Boost
Imported Data	
Exported Data	
Notes	
Train	
Variables	
∃Series Options	
N Iterations	50
Seed	12345
Shrinkage	0.1
Train Proportion	60
∃Splitting Rule	
Huber M-Regression	No
Maximum Branch	2
Maximum Depth	2
Minimum Categorical Size	5
Reuse Variable	1
Categorical Bins	30
Interval Bins	100
Missing Values	Use in search
Performance	Disk
∃Node	
Leaf Fraction	0.1
Number of Surrogate Rules	0
Split Size	.
∃Split Search	
Exhaustive	5000
Node Sample	20000
∃Subtree	
Assessment Measure	Average Square Error
Score	
Subseries	Best Assessment Value
Number of Iterations	1
Create H Statistic	No

results of a gradient boosting model. The subtree assessment plot (Fig. 7.3) shows
the average squared error in each successive tree that was created.

The train and validation trees show improvement in the average squared error
with each successive split up to iteration 16 with little improvement thereafter.
However, the model does not appear to be overfitting. The gradient boosting results
also show the variable importance (Fig. 7.4). SAS Enterprise Miner™ considers
both split-based variable importance and observation-based variable importance.

Fig. 7.3 Gradient boosting subtree assessment plot

Variable Name	Label	Number of Spitting Rules	Importance	Validation Importance	Ratio of Validation to Training Importance
Gender	Gender	20	1	1	1
Vehicle_Class	Vehicle_Class	21	0.895099	0.889983	0.994285
Income	Income	15	0.583727	0.267983	0.459089
LOG_Months_Since_Last_C..	Transformed: Months_Since..	11	0.423679	0.085476	0.201748
IMP_Education	Imputed: Education	9	0.368873	0.081645	0.221336
IMP_Outstanding_Balance	Imputed: Outstanding_Balan..	9	0.344613	0.074295	0.215589
State_Code	State_Code	7	0.318184	0.14031	0.440971
Employment_Status	Employment_Status	4	0.29461	0.387132	1.314048
Marital_Status	Marital_Status	5	0.279517	0.361026	1.291606
Vehicle_Size	Vehicle_Size	1	0.103768	0.089077	0.85842

Fig. 7.4 Gradient boosting variable importance

Split-based variable importance provides a list of the variables that contributed to the reduction in the residual for each split. In addition, it considers correlated variables by fully crediting each variable. The importance column is computed based upon the training data. In addition, observation-based variable importance which is not adjusted for correlated variables is also available. Variable importance can be an important explanatory feature when evaluating the inputs that had the most significant impact on a gradient boosting model.

When a gradient boosting model is applied to the automobile insurance claim data set, based upon an assessment of the average squared error, the results (Fig. 7.5) are very similar to the prior models using regression, decision trees, and neural networks.

7.3   Ensemble Models

An ensemble (Fig. 7.6) is a model of models, created from models that were independent of each other and that use the same target. An ensemble model is created by taking the predicted values from interval targets or the posterior probabilities from nominal or binary targets from the models that are input to the ensemble.

An ensemble model provides a new model that results from the best-fit models from a series of other predictive models usually, though not always, resulting in an even better-fit model. A common implementation of an ensemble model is to first evaluate the best-fit model among a variety of decision trees, neural networks, and regression models. From each group, the best-fit model is then input to the ensemble model. Other models, such as a gradient boosting, may also serve as input to an

Fit Statistics					
Target	Target Label	Fit Statistics	Statistics Label	Train	Validation
Fraudulent_Claim	Fraudulent_Claim	_NOBS_	Sum of Frequencies	2997	2002
Fraudulent_Claim	Fraudulent_Claim	_SUMW_	Sum of Case Weights Times Freq	5994	4004
Fraudulent_Claim	Fraudulent_Claim	_MISC_	Misclassification Rate	0.061061	0.061938
Fraudulent_Claim	Fraudulent_Claim	_MAX_	Maximum Absolute Error	0.976784	0.986259
Fraudulent_Claim	Fraudulent_Claim	_SSE_	Sum of Squared Errors	319.321	219.3095
Fraudulent_Claim	Fraudulent_Claim	_ASE_	Average Squared Error	0.053273	0.054773
Fraudulent_Claim	Fraudulent_Claim	_RASE_	Root Average Squared Error	0.23081	0.234035
Fraudulent_Claim	Fraudulent_Claim	_DIV_	Divisor for ASE	5994	4004
Fraudulent_Claim	Fraudulent_Claim	_DFT_	Total Degrees of Freedom	2997	

Fig. 7.5 Gradient boosting results

Fig. 7.6 Ensemble node

ensemble model. In SAS Enterprise Miner™ however, a random forest (described below) cannot serve as input to an ensemble. If there is no difference between the models that are input into the ensemble, then the ensemble will not generate an improved model. Therefore, you should always compare the individual model's performance to each other and the ensemble model performance to the individual models.

The ensemble node has few properties (Fig. 7.7). The most significant properties are the predicted values for interval targets or posterior possibilities for class targets.

The values supported by the properties include:

1. *Average*—Utilizes the average of the predicted values or posterior probabilities as the prediction for the ensemble model.
2. *Maximum*—Utilizes the maximum predicted value or posterior probability as the prediction for the ensemble model.
3. *Voting*—Utilized only for class targets. When this method is selected, the voting posterior probabilities property must be selected. Its values include:

   A. *Average* (default)—Uses the average posterior probabilities from only those models that predict the same target event. If a model does not predict that event, it is not included.
   B. *Proportion*—Calculates the posterior probability for a target variable from the proportion of models that predict the same target event. In this case, models that do not predict that event are still considered in the denominator of the resulting proportion.

When an ensemble node is applied to the automobile insurance claim data set using the best-fit decision tree, neural network, and regression nodes in combination with the gradient boosting, it results in an improved model based upon the results of the average squared error (Fig. 7.8). The models that were combined were not identical, but each represented the best of their group. In this way, an ensemble model can be thought of as an improvement of an improvement or (improvement^2).

Fig. 7.7 Ensemble
properties

Fig. 7.8 Ensemble results

7.4 Random Forests

A random forest is an ensemble model of multiple decision trees designed to support both regression and classification trees. In effect, it is a forest of decision trees. The purpose of combining multiple trees into a forest is to get a more accurate and stable prediction than a single decision tree that is capable of producing. A single decision tree is prone to new splits being generated as a result of small changes to the training data. By using multiple trees, the goal is to reduce the changes and produce a more stable model. SAS Enterprise Miner™ supports random forest modeling through its high-performance HP Forest node (Fig. 7.9).

The HP Forest node is designed to work with big data data sets. It uses the average of many trees to build a single tree model. This process is referred to as *bagging*. Both rows and columns of data are randomly sampled to determine where splits will occur, thus creating greater variation in the trees that are generated. Once the trees are constructed, the training data can be used to find the combination of splits that conclude with the most accurate result. This occurs by randomly selecting a group

Fig. 7.9 HP Forest node

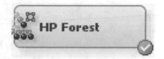

Target	Target Label	Fit Statistics	Statistics Label	Train		Validation	T:
Fraudulent_Claim	Fraudulent_Claim	_ASE_	Average Squared Error		0.054583		0.055366
Fraudulent_Claim	Fraudulent_Claim	_DIV_	Divisor for ASE		5994		4004
Fraudulent_Claim	Fraudulent_Claim	_MAX_	Maximum Absolute Error				0.970965
Fraudulent_Claim	Fraudulent_Claim	_NOBS_	Sum of Frequencies		2997		2002
Fraudulent_Claim	Fraudulent_Claim	_RASE_	Root Average Squared Error		0.23363		0.235299
Fraudulent_Claim	Fraudulent_Claim	_SSE_	Sum of Squared Errors		327.1717		221.6836
Fraudulent_Claim	Fraudulent_Claim	_DISF_	Frequency of Classified Cases		2997		2002
Fraudulent_Claim	Fraudulent_Claim	_MISC_	Misclassification Rate		0.061061		0.061938
Fraudulent_Claim	Fraudulent_Claim	_WRONG_	Number of Wrong Classifications		183		124

Fig. 7.10 Random forest results

of variables that predict our target variable. Random forests perform a split on only the variable that has the largest association with the target among the variables chosen at random within that iteration. The split is performed only on that variable within that iteration. This process continues until the tree is fully grown or you have reached the maximum number of iterations (set by the maximum number of trees property). A tree is fully grown when it has built a decision tree that contains one observation within each node uniquely explained by the decisions from the nodes that preceded it.

An advantage of random forests is that it works with both regression and classification trees so it can be used with targets whose role is binary, nominal, or interval. They are also less prone to overfitting than a single decision tree model. A disadvantage of a random forest is that they generally require more trees to improve their accuracy. This can result in increased run times, particularly when using very large data sets.

If we apply a random forest to the automobile insurance claim data set, the results show an average squared error (Fig. 7.10) in the relevant range of the other predictive models that have been utilized thus far.

The output of the random forest includes an iteration history (Fig. 7.11) which shows the results of the iterations of tree structure that were generated to get to the final result. This shows how quickly (or slowly) the model converges to the final result (i.e., the number of trees that it took to get to final fit statistics) as well as the complexity of the trees (based upon the number of leaves).

7.5   Memory-Based Reasoning

Memory-based reasoning (MBR) (Fig. 7.12) is a model that applies a k-nearest neighbor algorithm to provide an empirical classification method to compare cases to prior cases. It uses historical data to find records that are similar to current cases.

☐ Iteration History

Number of Trees	Number of Leaves	Average Square Error (Train)	Average Square Error (Out of Bag)	Average Square Error (Validate)	Misclassification Rate (Train)	Misclassification Rate (Out of Bag)	Misclassification Rate (Validate)	Log Loss (Train)	Log Loss (Out of Bag)	Log Loss (Validate)
1	2	0.0561	0.0545	0.0570	0.0611	0.0592	0.0619	0.212	0.206	0.244
2	5	0.0561	0.0584	0.0569	0.0611	0.0637	0.0619	0.212	0.219	0.243
3	9	0.0549	0.0573	0.0557	0.0611	0.0636	0.0619	0.203	0.219	0.235
4	12	0.0547	0.0560	0.0555	0.0611	0.0621	0.0619	0.204	0.215	0.207
5	14	0.0548	0.0565	0.0556	0.0611	0.0626	0.0619	0.205	0.211	0.208
6	17	0.0546	0.0553	0.0554	0.0611	0.0613	0.0619	0.204	0.207	0.207
7	19	0.0547	0.0555	0.0555	0.0611	0.0615	0.0619	0.205	0.208	0.208
8	22	0.0548	0.0554	0.0556	0.0611	0.0611	0.0619	0.205	0.208	0.208
9	26	0.0546	0.0554	0.0555	0.0611	0.0614	0.0619	0.203	0.207	0.206
10	31	0.0546	0.0555	0.0554	0.0611	0.0615	0.0619	0.203	0.208	0.206
11	33	0.0546	0.0554	0.0555	0.0611	0.0614	0.0619	0.204	0.208	0.207
12	36	0.0546	0.0552	0.0554	0.0611	0.0612	0.0619	0.203	0.207	0.207
13	38	0.0546	0.0552	0.0555	0.0611	0.0611	0.0619	0.204	0.207	0.207
14	40	0.0547	0.0552	0.0555	0.0611	0.0611	0.0619	0.204	0.208	0.208
15	44	0.0546	0.0550	0.0554	0.0611	0.0611	0.0619	0.203	0.206	0.207
16	47	0.0546	0.0550	0.0554	0.0611	0.0611	0.0619	0.203	0.206	0.207
17	51	0.0546	0.0549	0.0554	0.0611	0.0611	0.0619	0.203	0.206	0.206
18	53	0.0546	0.0549	0.0554	0.0611	0.0611	0.0619	0.204	0.206	0.207
19	56	0.0546	0.0549	0.0554	0.0611	0.0611	0.0619	0.204	0.206	0.207
20	58	0.0546	0.0549	0.0554	0.0611	0.0611	0.0619	0.204	0.206	0.207
21	61	0.0546	0.0550	0.0555	0.0611	0.0611	0.0619	0.204	0.206	0.207
22	65	0.0546	0.0549	0.0554	0.0611	0.0611	0.0619	0.203	0.206	0.206
23	68	0.0546	0.0549	0.0554	0.0611	0.0611	0.0619	0.203	0.206	0.207
24	70	0.0546	0.0549	0.0554	0.0611	0.0611	0.0619	0.204	0.206	0.207
25	73	0.0546	0.0549	0.0554	0.0611	0.0611	0.0619	0.204	0.206	0.207

Fig. 7.11 Random forest iteration history

Fig. 7.12 Memory-based
reasoning

Most often these will not result in exact matches but will seek to find the historical cases that most closely approximate the current case. To accomplish this, a training data set and probe are used. A probe represents one value for each variable. The
distance between the probe and the observation is calculated. Within SAS Enterprise
Miner™, the k-nearest neighbor algorithm used to calculate this distance is
Euclidean distance. The *nearest neighbor* is the shortest distance between the probe
and the observation. For binary or nominal target variables, posterior probabilities
are used to determine the k-nearest neighbor. Memory-based reasoning models are
based upon the assumptions that the data is numeric, standardized, and orthogonal
between the probe and observation. Principal component analysis is used to transform the raw variables to ensure they are standardized and orthogonal. This results
in transforming variables that are possibly correlated into linearly uncorrelated variables; as a result it is sensitive to the scale of the original variables. This can be
accomplished with a principal components node prior to using memory-based reasoning. Since the input variables must be numeric, categorical variables need to be
transformed into numeric values. In addition, it may be necessary to reduce the
number of categories to reduce the sensitivity to the scale and thus reduce the possibility of overfitting the training data. This can be accomplished with a
Transform node.

The memory-based reasoning node requires that the data contain only one target
variable. The target variable can be nominal, binary, or interval. An ordinal target is
not supported; however, it can be changed to an interval variable. If there is more

than one target variable on the input data source, prior to the memory-based reasoning node, this must be reduced to a single target.

The memory-based reasoning node contains several properties that directly influence how the node performs (Fig. 7.13). These include:

1. *Method*—Permits one of two possible data representation methods used to store the training data observations to then subsequently retrieve the nearest neighbors. These include:

   A. *RD-Tree*—This is the default setting. The Reduced Dimensionality Tree method usually works better than the Scan method for data sets that contain dimensionality up to 100 (i.e., it will generally perform better when you have a relatively small number of variables). The RD-Tree partitions the data set into smaller subsets by creating a binary tree in memory. Splits usually occur along the median value for the variable that has the greatest variation for the observations.

   B. *Scan*—Retrieves a k-nearest neighbor by calculating its Euclidean distance from a probe to an observation. This method generally outperforms the RD-Tree when dealing with data sets that contain a large number of variables (e.g., greater than 100).

2. *Number of Neighbors*—Specifies the number of nearest neighbors that you want to use to categorize or predict observations (i.e., the k in k-nearest). Usually the lower the number of neighbors, the higher the error that will result.

3. *Epsilon*—Used with the RD-Tree method to specify a distance for an approximate nearest neighbor search. Using this when the data has high dimensionality can result in a significant performance improvement as it will result in searches terminating earlier. This property does not apply to the Scan method.

Fig. 7.13 Memory-based reasoning properties

. Property	Value	
General		
Node ID	MBR	▲
Imported Data		▣
Exported Data		▣
Notes		▣
Train		≡
Variables		▣
Method	RD-Tree	▼
Number of Neighbors	16	
Epsilon	0.0	
Number of Buckets	8	
Weighted	Yes	
Create Nodes	No	
Create Neighbor Variables	Yes	▼

4. *Number of Buckets*—Used with the RD-Tree method to specify the maximum number of buckets permitted for a leaf node. Once this number is exceeded, the leaf will split into a branch with two new leaves. The minimum value is 2 and the default is 8. This influences the size and shape of the resulting binary tree. This property does not apply to the Scan method.
5. *Weighted*—Used with an interval target variable to weigh each input variable by the absolute value of its correlation to the target variable when set to Yes (the default value).
6. *Create Nodes*—Creates a variable called _NNODES_ to the memory-based reasoning node output data set when this property is set to Yes. The _NNODES_ variable may be useful when performing a point comparison because it shows the number of point comparisons that were performed during node calculations. The default setting is No.
7. *Create Neighbor Variables*—Writes the nearest neighbors for each observation in the memory-based reasoning node output score data set. This permits you to view which records were closely associated with each observation. The variables for the nearest neighbors are named $\_N_1$, $\_N_2$, ..., $\_N_k$, where k is the number of nearest neighbors specified in the number of neighbor properties. If an ID variable is defined, the ID value is contained in the generated variable. If there is no ID variable, then the relative record number in the data set is contained in the generated variable. The default setting is Yes.

To apply a memory-based reasoning node to the automobile insurance claim data set, we would first utilize a principal component (Fig. 7.14) and then connect it to the memory-based reasoning node. The input to the principal components node could be the output of the Transform Variables node (or any node thereafter).

Using the default method (RD-Tree), the results show an average squared error (Fig. 7.15) in the relevant range of the other predictive models that have been utilized thus far, though not as good as the ensemble model results.

7.6 Two-Stage Model

The two-stage node (Fig. 7.16) permits the modeling of two target variables simultaneously. One of the target variables is a class variable, and the other is an interval variable that usually represents the value associated with the level of the class variable. In many business problems, it is common to not only want to know the likelihood that an event will occur but the impact of the event. In those cases, a two-stage model analysis may be appropriate.

Fig. 7.14 Memory-based reasoning diagram

Target	Target Label	Fit Statistics	Statistics Label	Train	Validation	
Fraudulent_Claim	Fraudulent_Claim	_NW_	Number of Estimated Weights	10		
Fraudulent_Claim	Fraudulent_Claim	_NOBS_	Sum of Frequencies	2997	2002	
Fraudulent_Claim	Fraudulent_Claim	_SUMW_	Sum of Case Weights Times Freq	5994	4004	
Fraudulent_Claim	Fraudulent_Claim	_DFT_	Total Degrees of Freedom	2997		
Fraudulent_Claim	Fraudulent_Claim	_DFM_	Model Degrees of Freedom	10		
Fraudulent_Claim	Fraudulent_Claim	_DFE_	Degrees of Freedom for Error	2987		
Fraudulent_Claim	Fraudulent_Claim	_ASE_	Average Squared Error	0.051544	0.057439	
Fraudulent_Claim	Fraudulent_Claim	_RASE_	Root Average Squared Error	0.227032	0.239664	
Fraudulent_Claim	Fraudulent_Claim	_DIV_	Divisor for ASE	5994	4004	
Fraudulent_Claim	Fraudulent_Claim	_SSE_	Sum of Squared Errors	308.9531	229.9844	
Fraudulent_Claim	Fraudulent_Claim	_MSE_	Mean Squared Error	0.051716	0.057439	
Fraudulent_Claim	Fraudulent_Claim	_RMSE_	Root Mean Squared Error	0.227412	0.239664	
Fraudulent_Claim	Fraudulent_Claim	_AVERR_	Average Error Function	0.178845	0.242298	
Fraudulent_Claim	Fraudulent_Claim	_ERR_	Error Function	1071.996	970.1616	
Fraudulent_Claim	Fraudulent_Claim	_MAX_	Maximum Absolute Error	0.9375	1	
Fraudulent_Claim	Fraudulent_Claim	_FPE_	Final Prediction Error	0.051889		
Fraudulent_Claim	Fraudulent_Claim	_RFPE_	Root Final Prediction Error	0.227791		
Fraudulent_Claim	Fraudulent_Claim	_AIC_	Akaike's Information Criterion	1091.996		
Fraudulent_Claim	Fraudulent_Claim	_SBC_	Schwarz's Bayesian Criterion	1152.05		
Fraudulent_Claim	Fraudulent_Claim	_MISC_	Misclassification Rate	0.061061	0.061938	
Fraudulent_Claim	Fraudulent_Claim	_WRONG_	Number of Wrong Classifications	183	124	

Fig. 7.15 Memory-based reasoning results

Fig. 7.16 Two-stage node

The two-stage node contains several properties that directly influence how the node performs (Fig. 7.17).

The Model-Type property is central to the performance of this model. It permits one of two possible values: sequential or concurrent. Sequential modeling (the default) creates a categorical prediction variable from the class target. It then uses that to model the interval target. The transfer function and filter option control how the class target model will be applied to the interval target model. The prediction of the interval target is optionally adjusted by the posterior possibilities of the class target by setting the bias property.

When concurrent modeling is utilized, a neural network is used to model the class and interval targets at the same time (Scientific Books 2015).

7.7   Comparing Predictive Models

Predictive analytics consists of a variety of different techniques. In addition, when analyzing a business problem, it is common to try not only multiple techniques but also multiple different models of the same technique. This is particularly true when you do more exploratory analysis because you are not that familiar with the subject data. This can result in diagrams that grow in terms of the number of nodes and complexity of the process. If we have two or more predictive models, it is critical to have a means to be able to compare them to determine which model to use. The Model Comparison node (Fig. 7.18) provides a common framework for comparing

Fig. 7.17 Two-stage node properties

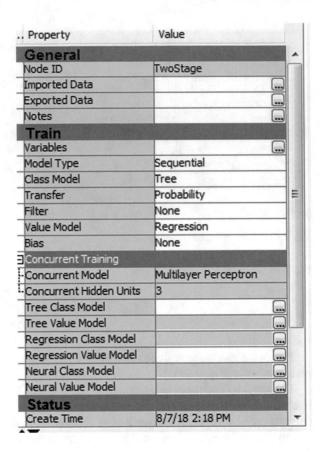

models and predictions from any of the modeling tools (such as regression, decision tree, and neural networks).

The model comparison is based on standard model fit statistics as well as potential expected and actual profits or losses that would result from implementing the model. The node produces charts that help describe the usefulness of the model: lift, profit, return on investment, receiver operating curves, diagnostic charts, and threshold-based charts. Multiple Model Comparison nodes may be used within a single predictive analytics diagram. For example, a model comparison may be used to determine which is the best fit of several decision tree models. There may be another model comparison to determine which is the best fit of several neural network models. Both of these model comparisons may then either be input to a third model comparison to determine which is the best overall model, or they may be input to an ensemble model or both (Fig. 7.19).

There are several measures that can be used to determine the best model out of a group of two or more models. Assessment statistics are computed for the training and validation data sets, as well as the test data set when present.

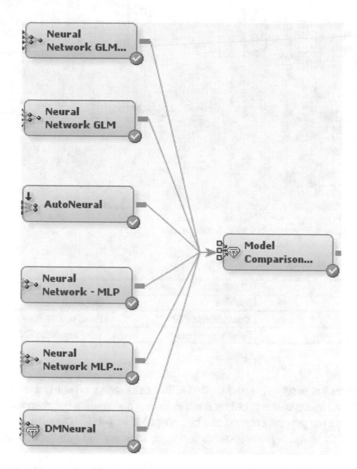

Fig. 7.18 Model comparison

7.7.1   Evaluating Fit Statistics: Which Model Do We Use?

SAS Enterprise Miner™ supports 14 different selection statistics to compare model performance. Although some are dependent upon the type of target variable that is specified, most are supported by a variety of target types. The following reviews these measures, beginning with the most widely used measures.

Misclassification rate is one of the most widely used selection statistics, particularly when the target value is binary. It measures the proportion of misclassified data against the total data. When comparing a predicted value to an actual (observed) value, there is one of four possible outcomes (Table 7.1).

The misclassification rate therefore is

$$MR = (FP + FN) / (TP + FP + FN + TN)$$

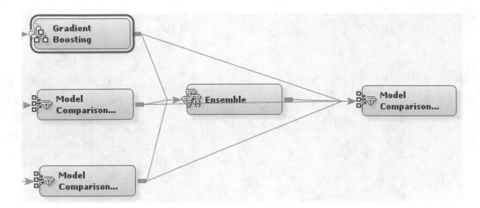

Fig. 7.19 Model comparison of a model comparison

Table 7.1 Data accuracy (Gonem 2007)

	Observed	
Predicted	Positive	Negative
Positive	True positive (TP)	False positive (FP)
Negative	False negative (FN)	True negative (TN)

It presents the percentage of data that either exhibited a type 1 (false-positive) or type II (false-negative) error, but by itself does not distinguish between the two. The *accuracy rate* (or percent correct) can be computed as 1-MR. Therefore, when comparing models based on misclassification rate, look for the smallest misclassification rate.

The true positive rate, which is also referred to as *sensitivity* , is calculated as

$$TPR = TP / (TP + FN).$$

The true negative rate, which is also known as *specificity*, is calculated as

$$TNR = TN / (TN + FP).$$

Sensitivity and specificity are used to generate a graph referred to as the *receiver operating characteristic* (ROC) curve. The *x*-axis of the ROC curve is 1-specificity, also referred to as the false-positive rate or *fall out* (Fig. 7.20 shows an example from the model comparison of the automobile insurance claim data set example).

The area under the curve is called the concordance statistic (C-statistic) and represents the goodness of fit for the binary outcomes. In other words, it represents the ability of the model to accurately predict the data. From this a ROC index is generated. The ROC index is the percent of concordant cases plus one-half the number of tried cases. A pair of cases is concordant when the primary outcome has a higher

Fig. 7.20   ROC curve

Fig. 7.21   Cumulative lift model comparison

rank than any secondary outcome. The model with the largest area under the curve (i.e., the highest ROC index) is considered the best fit when this statistic is used for model comparison. If the ROC index is <6, it is considered to be a weak index. If the ROC index is >7, then the index is considered to be strong.

The *Gini coefficient* is calculated from ROC index. SAS Enterprise Miner™ calculates the Gini coefficient = 2 × (ROC index—0.5). It is an alternative to the ROC when ranking binary predictions. It was originally developed by Corrado Gini (1921) to be used as a technique to measure income distribution.

The *cumulative lift* is a measure of the effect of the predictive model by evaluating the results achieved with and without the model. In other words, it provides a measure to evaluate the performance of a model versus random guessing (Fig. 7.21 shows a cumulative lift chart for the model comparison of the automobile insurance claim example).

The data is sorted in descending order by the confidence the model assigns to the prediction. The *x*-axis reflects the results as a percentage of the overall data. When using cumulative lift as a model comparison, the higher the lift, the stronger the model.

Although *average squared error* and *mean squared error* are both possible as a selection statistic for model comparison, average squared error is more commonly used. Mean squared error is used in linear models. It is the sum of squared errors divided by the degrees of freedom for error. The degrees of freedom for error are calculated as the number of cases minus the number of weights in a model. Although the result is considered an unbiased estimate of the population noise variance, mean squared error fails as a selection statistic for decision trees and neural networks. In both cases there is a known unbiased estimator, and the degrees of freedom for error for neural networks are often negative. The average squared error is a more reliable measure in these cases. Instead of the degrees of freedom for error, it is the sum of squared error divided by the number of cases. It can be used with targets that are categorical or interval. When used for model comparison, the lowest error is considered the best-fit model.

The *Kolmogorov-Smirnov statistic* (often referred to as the KS statistic) describes the ability of a model to separate primary outcomes from secondary outcomes. When used as a model comparison statistic, the highest Kolmogorov-Smirnov statistic value is considered the best-fit model.

The *Akaike's information criterion* is an estimator of relative quality among two or more models. When used as a model selection criterion, the smallest index indicates the preferred model (Akaike 1974). This criterion is often used when evaluating two-stage node models.

7.8   Using Historical Data to Predict the Future: Scoring

Though it is essential to use historical data to build predictive models, the purpose of predictive modeling is to be able to put the models into action. When the final predictive model has been completed, it provides the best fit based upon the analysis of the historical data. For this to be of value to an organization, it then needs to be applied to new business activity. This is where the power of predictive analytics is demonstrated. This process is referred to as *scoring*. Scoring applies an existing predictive model to a new transaction data set for the purpose of generating either a probability or expected value for a target variable outcome. This depends upon the role (data type) of the target variable. If the target is binary or nominal, then a probability will result. If the target is an interval, then an expected value will be calculated. To perform this process, two inputs are required to the Score node (Fig. 7.22). One is the predictive model and the other is a score data set.

A score data set represents new transactions to be evaluated against a predictive model. These are transactions where the outcome is not yet known. When the data

Fig. 7.22 Scoring new data

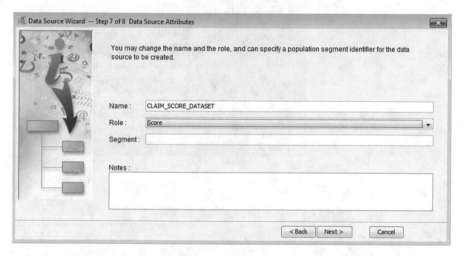

Fig. 7.23 Creating a score data set

source is created, it must be identified as a score data set. This is done in Step 7 of the Create Data Source process (Fig. 7.23).

With these two inputs, the Score node is ready to be processed. Once it is run, the output can be evaluated. To do so, select the exported data property from the Score node (Fig. 7.24) by clicking on the three ellipses.

This will open a window that presents all of the data sets available to view. Highlight the score data set (Fig. 7.25) and select Explore.

7.8.1 Analyzing and Reporting Results

Explore presents the data with additional variables added as a result of the scoring process. For a binary target (such as in the automobile insurance claim data set example), two predictor variables are created, each one representing one possible outcome for the target variable. In this example, they are Predicted Fraudulent_ Claim = N and Predicted Fraudulent_Claim = Y (Fig. 7.26).

By selecting any column heading, the results will be sorted based upon that column. In this case selecting Predicted Fraudulent_Claim = Y will sort the claim

Fig. 7.24 Accessing a
scored data set

Fig. 7.25 Exported data

records based upon their highest probability of fraud. The claims most likely to be
fraudulent could then be further investigated. Note that the results that are typically
achieved in most cases are not perfect probabilities (i.e., we don't typically see
probabilities of 100% that a claim is fraudulent, etc.). However, even with a relatively low probability, an organization can utilize those results to make decisions
that might not otherwise have been apparent to them.

	Claimant__Number		Predicted: Fraudulent_Claim=N	Predicted: Fraudulent_Claim=Y ▼			Predictio...
...N	6897	...N N	0.813541	0.186459	...N	...N	
...N	6727	...N N	0.815313	0.184687	...N	...N	
...N	7649	...N N	0.816998	0.183002	...N	...N	
...N	7532	...N N	0.81766	0.18234	...N	...N	
...Y	7940	...N N	0.819059	0.180941	...N	...N	
...N	7944	...N N	0.819523	0.180477	...N	...N	
...N	6106	...N N	0.819897	0.180103	...N	...N	
...N	6221	...N N	0.820191	0.179809	...N	...N	
...N	7046	...N N	0.821312	0.178688	...N	...N	
...N	6951	...N N	0.821738	0.178262	...N	...N	
...N	6462	...N N	0.82268	0.17732	...N	...N	
...N	7422	...N N	0.82322	0.17678	...N	...N	
...N	7609	...N N	0.824515	0.175485	...N	...N	
...N	6915	...N N	0.824543	0.175457	...N	...N	
...N	7503	...N N	0.824571	0.175429	...N	...N	
...N	7985	...N N	0.825071	0.174929	...N	...N	
...N	7771	...N N	0.825823	0.174177	...N	...N	
...N	6062	...N N	0.825845	0.174155	...N	...N	
...N	7535	...N N	0.82623	0.17377	...N	...N	
...N	6264	...N N	0.827098	0.172902	...N	...N	
...N	7986	...N N	0.827108	0.172892	...N	...N	
...N	7381	...N N	0.827643	0.172357	...N	...N	
...N	6294	...N N	0.827679	0.172321	...N	...N	
...N	6874	...N N	0.827712	0.172288	...N	...N	
...N	7160	...N N	0.827946	0.172054	...N	...N	
...N	7079	...N N	0.828096	0.171904	...N	...N	
...N	7574	...N N	0.828121	0.171879	...N	...N	
N	6205	N N	0.828177	0.171823	N	N	

Fig. 7.26 Exploring the scored results. The output results have been condensed to show the probabilities by Claimant_number

Fig. 7.27 Save data

7.8.2   Save Data Node

Once you have completed scoring the data, the output of the Score node is a temporary file. In many cases, it is important to keep the data set for future use. This can be done using a Save Data node (Fig. 7.27).

There are two key properties (Fig. 7.28) that must be set. These are the File Format and SAS Library Name.

The File Format property specifies the file type of the output data set. The default is a SAS-formatted file; however, this can be changed to:

1. JMP—used by the SAS JMP software
2. TXT—text file format
3. CSV—comma separated file format
4. XLSX—Microsoft Excel worksheet format

The SAS Library Name property specifies the output directory where the file will be stored. The default value is the SAS library that was defined for the diagram.

Fig. 7.28 Save Data
properties

Fig. 7.29 Reporter node

7.8.3   Reporter Node

The Reporter node (Fig. 7.29) provides a mechanism to produce a comprehensive
report of all of the nodes that were executed in a process flow model up to the
reporter node.

When listed as the last node in a process flow, it will provide a report of the entire
model beginning with a diagram of the model and followed by a report for the first
node and each subsequent node. The default is a report in a PDF format; this can be
changed to an RTF format if needed. There are eight Summary Report properties
(Fig. 7.30) that define the output reports that will be created.

The Summary Report properties include:

1. Basic Reports—These include a gain chart, variable importance, and ROC chart
   (available only for categorical targets).
2. Summarization—Includes a Summary Data section of the input data sources and
   target variables.
3. Variable Ranking—Includes variable rankings of all of the input variables.
4. Classification Matrix—Includes a classification matrix that identifies the true
   positives, true negatives, false positives, and false negatives in the output report.
5. Cross Tabs—Includes a cross tab report in the output report if the target variable
   is nominal.

Fig. 7.30 Report
properties

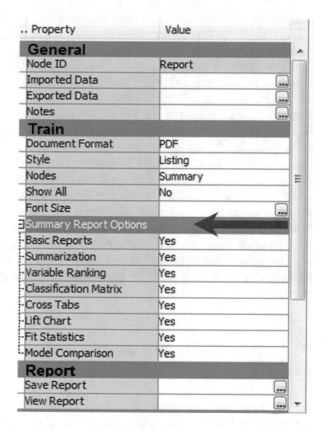

6. List Chart—Includes a cumulative lift chart in the output report.
7. Fit Statistics—Includes a table for the training and validation fit statistics that were used to evaluate the model.
8. Model Comparison—Includes a model comparison table to show how the best-fit model was selected. This is key to describing the results of a predictive analytics analysis.

7.9   The Importance of Predictive Analytics

The value of predictive analytics does not have to be measured in absolutes. Organizations can achieve substantial benefits from identifying more likely outcomes. Historical data will only generate absolute probabilities when there is a perfect pattern to the data. This is almost never the case. However, a vast amount of historical data can now be analyzed to uncover possibilities that were previously unknown. In this way, we can finally make use of the tremendous volumes of data that many organizations have been accumulating for years. To make accurate predictions, we must have good data. This is not limited to the data merely being clean; it must also contain the values that both represent and can be used to predict what we want to study.

For well over a decade, many organizations have been collecting data to build customer profiles. Some effectively utilize the profile to build loyalty with customers by accurately identifying the customer characteristics that predict their future behavior (often intention to purchase). How data is collected and managed continues to be an issue. Views on data privacy and security have changed over time so it is important to take that into account when determining which data is acceptable to collect and use and which is not. Davenport in a Harvard Business Review report (2014) points outs that every model has assumptions built into it. It is important to understand what these underlying assumptions and monitor if they remain true. The primary assumption of predictive analytics is that historical data is representative of future events. When this assumption changes and is no longer true, new actions must be taken. Either predictive analytics is no longer appropriate, or new data that is more representative of the current events must be used. Business conditions change over time; they must be monitored to adjust predictive models where necessary.

7.9.1   What Should We Expect for Predictive Analytics in the Future?

We continue to see new applications of predictive analytics. It is a fast-growing field where new and exciting uses will continue to emerge. Siegel (2018) describes some potential areas where predictive analytics will impact our daily lives in the not-too-distant future. Some of these examples include:

1. *Anti-theft*—Your automobile will automatically scan and authenticate the driver based on biometric data.
2. *Entertainment*—Your entertainment choices will be personalized to your preferences and automatically selected for you.
3. *Traffic*—Your navigator automatically suggests new routes due to delays. How far are we from it automatically rerouting the vehicle without the need for driver intervention?

The Internet of Things is upon us, and it is no longer a view of things to come. Consumers have been embracing the technology, and more of it continues to become interconnected. As we look to the future, we see the potential to continue to integrate predictive analytics with this technology. It won't be long before your refrigerator will send a message to your smartphone to remind you to buy milk because you are almost out of it.

The greater the benefits organizations receive from predictive analytics, the more they will invest in it. For example, Edwards (2018) reported that a 2017 Society of

Actuaries report on the healthcare industry trends in predictive analytics reported that a 15% or more savings will occur over the next 5 years as a result of predictive analytics. This was from a survey of healthcare executives at organizations already using predictive analytics. An additional 26% of those surveyed believed the savings would be in the 25% range. This can result in revolutionary change in healthcare, particularly if these savings can be reinvested in new treatments and cures.

7.10   Chapter Summary

The Big 3 in predictive analytics are the best-known and most widely used techniques; however new methods to analyze data continue to be developed. Gradient boosting provides a series of decision trees that resample data to correct inaccuracies in the prior trees by fitting a residual of the prediction from the earlier trees.

An ensemble model is a model of models. It is generally implemented by taking the best of the two or more predictive models, combining them to form what is usually an even stronger model. Often this consists of a decision tree and a neural network (at a minimum); however almost any type of model can be combined. The ensemble node permits the user to specify if the models will be averaged or utilized the maximum predicted value or posterior probability. For class variables, a voting method is also supported. Random forests utilize multiple decision trees that support both regression and classification trees. The purpose of multiple trees is to provide a more stable predictor than a single decision tree. Random forests use a process called bagging to create a single decision tree from the average of many trees. Memory-based reasoning applies the k-nearest neighbor algorithm to use historical data to closely approximate new cases using one of two possible methods: Reduced Dimensionality Trees (RD-Trees) or the Scan method. The two-stage model supports simultaneous analysis of a class and interval target variable. This is useful when we want to know not only the probability that an event will occur but also the possible impact. For business problems, the possible impact is often measured in dollars representing either possible losses or potential increases to revenue.

It's rare in predictive modeling that we only create only a single model. With so many techniques available after all, why limit our analysis to one choice? As a result, though, a method to easily asses which model is the best fit for the particular problem is required. Model comparison provides a means to use several different selection criteria against multiple predictive models to show which model is the best fit. This is an important step. After all, it is the best-fit model that should be used to analyze current business activity. Some of the most common statistics for evaluating predictive models include the misclassification rate, average squared error, ROC index, and cumulative lift. The business problem to be solved and the data type of the target variable impact the measure to be used.

Scoring is the term used for the process that gives value and meaning to predictive modeling. Once a predictive model has been constructed, it should be used. Scoring combines the predictive model with new cases to either generate posterior probabilities for these new cases or to predict values based upon the target variable. The value of big data is utilized to generate effective predictive models and scoring and then enables an organization to learn from that data as they move forward to seek out new opportunities or boldly go where no organization has gone before.

Discussion Questions

1. Identify at least three sources of data an organization might use to find information about prospective new customers. From that data, how might they use predictive analytics to determine which individuals are most likely to become a customer.
2. Describe at least three ways in which predictive analytics can be used within the healthcare industry.
3. Discuss the differences between fit statistics and provide an example of when it is appropriate to utilize each one in determining which predictive model is the best fit.
4. What properties could be changed to improve the results of the automobile insurance claim data set example that resulted from using the memory-based reasoning model? Discuss the impact on the model that resulted from the changes.
5. Describe one other selection statistic, and discuss why it would or would not be appropriate to use with the automobile insurance claim data set example.
6. Discuss how predictive analytics can be used in the future to provide new products or services.

References

Akaike H (1974) A new look at the statistical model identification. IEEE Trans Automat Contr 19(6):716–723
Browniee J (2016) A gentle introduction to XGBoost for applied machine learning. https://machinelearningmastery.com/gentle-introduction-xgboost-applied-machine-learning/
Davenport T (2014) A predictive analytics primer, within predictive analytics in practise. Harvard Business Review report, Harvard Business Publishing
Edwards J (2018) What is predictive analytics? Transforming Data into Future Insights. https://www.cio.com/article/3273114/predictive-analytics/what-is-predictive-analytics-transforming-data-into-future-insights.html
Gini C (1921) Measurement of Inequality of Incomes. Econ J 31(121):124–126
Gonem M (2007) Analyzing receiver operating curves with SAS®. SAS Publishing, Cary
Scientific Books (2015) Data mining techniques, predictive models with SAS Enterprise. Create Space Independent Publishing, Miner
Siegel E (2018) The future of prediction: predictive analytics in 2010. https://bigthink.com/experts-corner/the-future-of-prediction-predictive-analytics-in-2020

Chapter 8
Finding Associations in Data Through Cluster Analysis

Chapter Learning Objectives
1. Identify and explain hierarchical cluster techniques.
2. Identify and explain non-hierarchical cluster techniques.
3. Use SAS Enterprise Miner™ to cluster and analyze data.

Clustering is a technique of grouping similar observations into smaller groups within the larger population. The resulting groups should be homogeneous, with each member of the cluster having more in common with members of the same cluster than with members of the other clusters. Cluster analysis is an exploratory data analysis tool which aims at sorting different objects into groups in a way to maximize the degree of association between objects in the same cluster. In this chapter several clustering techniques are explained. A simple example is used to explain the different clustering methods. Finally, clustering is applied to a subset of the automobile insurance data set.

8.1 Applications and Uses of Cluster Analysis

Cluster analysis has a widespread application in business analytics. One of the questions businesses often face is how to organize huge amounts of available data into meaningful structures. Smaller homogeneous groups of data can be easier to work with rather than a single large heterogeneous data set. Clustering can be used in:

- *Target marketing* to group customers according to different criteria such as the number of purchases, amount spent, or days since the last purchase. The resulting clusters allow companies to target customers with different offers and incentives individualized for each of the different clusters.

- *Estimating* the value of *real estate* property. Clustering can be applied to identify groups of homes based on the features of the house and its location.
- *Fraud detection* to detect outliers or anomalies. Clustering can be used to identify transactions that are outside of the normal business process.
- *Healthcare* to discover patterns such as if certain geographical locations are associated with high or low levels of illnesses.

In summary, cluster analysis is a useful technique to explore patterns or associations within data and is broadly used in many applications.

8.2 Types of Clustering Techniques

There are several different clustering techniques available, and the results of each technique may vary. The selection of a specific technique may depend on the clustering objective, the desired results, the technology available, and the size of the data set. Keep in mind that there are no right or wrong clustering solutions, but there are solutions that are more meaningful and valuable for specific projects.

Clustering techniques can be divided into two categories, hierarchical and non-hierarchical. *Hierarchical clustering analysis* (HCA) is an unsupervised technique that produces a tree-like structure of clusters showing the relationship among the observations. An unsupervised technique is one in which there is no predetermined target variable(s) within the data set. The clusters are defined in a predefined order based upon the algorithm that determines the similarity of characteristics from top to bottom. Hierarchical clustering is often applied to hierarchical data, such as social network data, web page organization, and taxonomies. Hierarchical methods when compared to non-hierarchical methods tend to be much more computationally expensive because all observations are compared before the next clustering step. Hierarchical clusters are generally easier to interpret.

Non-hierarchical clustering analysis (NHCA) techniques create new clusters by merging or splitting clusters. The clusters do not have a hierarchical relationship. The clusters are determined by maximizing or minimizing a specific evaluation criterion. The starting observation (seed) or cluster can impact the results. Non-hierarchical clustering is more reliable and more commonly used as it is less prone to error.

8.3 Hierarchical Clustering

Hierarchical clustering forms a tree-like structure of clusters with all the observations merged into one cluster at the top of the tree. Hierarchical clustering algorithms are either *agglomerative* or *divisive*. The agglomerative approach is a bottom-up approach where each observation is its own cluster, and then as you traverse the hierarchy, the two most similar clusters are joined. The divisive approach is a top-down approach where all the observations begin as one cluster, and then as you traverse down the hierarchy, each cluster is divided into more homogeneous clusters.

8.3.1 Agglomerative Clustering

Agglomerative clustering also known as AGNES (agglomerative nesting) is the most widely used type of hierarchical clustering. The steps for the agglomerative clustering are:

1. Assign each observation as its own single cluster.
2. Calculate the distance measurements and the corresponding distance matrix.
3. Determine and apply the linkage criteria to join the clusters.
4. Recalculate the distance matrix.
5. Repeat the process until every observation becomes one cluster.

There are several distance measurements that can be used. More popular measurements and their formulas are given in Table 8.1. The Euclidean distance treats each variable equally. Mahalanobis distance developed by P.C. Mahalanobis (1936) accounts for the variance of each variable and the covariance between variables. It considers the scale and distribution of the data.

There are also several linkage criteria that can be used including single linkage, complete linkage, average linkage, centroid, and Ward's. The mathematical formulas for the linkages are given in Table 8.2. In single linkage all the distances between the items in the two clusters are calculated, and the smallest distance is taken. Complete linkage takes the maximum distance. The average linkage developed by Sokal and Michener (1958) takes the average distance between each point in one

Table 8.1 Distance measurements

Measurement	General formula
Euclidean	$\lvert x - y \rvert = \sqrt{\sum_{i=1}^{n} \lvert x_i - y_i \rvert^2}$
Manhattan	$\lvert x - y \rvert = \sum_{i=1}^{n} \lvert x_i - y_i \rvert$
Mahalanobis	$D^2 = (x - m)^T C^{-1} (x - m)$ Where x is the vector of observation (row in a data set) m is the mean values of independent variables (mean of each column) C^{-1} is the inverse covariance matrix of independent variables (Prabhakaran 2019)

Table 8.2 Linkage criteria

Linkage method	Formula
Single linkage	$D_{\min}(C_i,C_j) = \min\limits_{x \in C_i,\, y \in C_j} d(x,y)$
Complete linkage	$D_{\max}(C_i,C_j) = \max\limits_{x \in C_i,\, y \in C_j} d(x,y)$
Average linkage	$\dfrac{1}{C_i C_j} \sum\limits_{x \in C_i,\, y \in C_j} d(x,y)$
Centroid	$D_{\text{means}}(C_i, C_j) = \|d(\mu_i, \mu_j)\|$
Ward's	Min (error sum of squares) $= \sum\limits_{i} \sum\limits_{j} \left(X_{ij} - \overline{X}_i\right)^2$

cluster to every point in the other cluster. The average linkage tends to join clusters with small variances and avoids extreme large or tightly compact clusters. The centroid method utilizes the squared Euclidean distance between the means. This method is better at handling outliers, but generally does not perform as well as Ward's or average. Ward's method criterion minimizes the total within-cluster variance. It tends to join clusters with few observations. It also tends to produce homogeneous clusters that are spherical in shape.

Let's look at an example of single linkage. Consider a simple marketing segmentation example where you are looking to segment your customers by the number of items purchased and customer income. The data and the plot of the data are given in Table 8.3 and Fig. 8.1.

Each observation is considered a cluster; so at the start, there are six clusters. Using the Euclidean method, as explained in Chap. 3, the distances between each cluster is calculated and placed in a matrix as shown in Table 8.4. In each iteration the closest pair of clusters is determined and then grouped together to form one cluster. Looking at Table 8.4, clusters 3 and 5 are the closest with a minimum distance of 1.00 and will be merged.

The next step is to recalculate the distance matrix. There are now five clusters (1), (2), (3,5), (4), and (6). To determine the minimum distance from the merged cluster to the other clusters, you need to calculate the minimum distance from each observation in the cluster to every other cluster. For example,

$$D(C_1,C_{3,5}) = \min\left((C_1,C_3),(C_1,C_5)\right) = \min(3.16,3.61) = 3.16$$

Table 8.3 Customer income and items purchased data

Observation	Income ($10 K)	Number of items purchased
#1	8	5
#2	7	4
#3	5	4
#4	4	1
#5	5	3
#6	9	2

Fig. 8.1 Customer income and items purchased plot

the recalculated distance matrix given in Table 8.5. The process repeats. Clusters 1 and 2 are the closest with a minimum distance of 1.41 and will then be merged.

A third distance matrix is calculated (Table 8.6), and the results indicate to merge the joined cluster (1,2) to the joined (3,5) cluster. The process continues until all the clusters are joined into one large cluster (Table 8.7). The resulting clusters are shown in the chart and corresponding dendrogram in Figs. 8.2 and 8.3. The dendrogram provides information on the order the clusters were joined, and the height indicates the distance between the clusters.

To illustrate further, let's look at the sample example but use the average linkage to determine the clusters. The initial distance matrix shown in Table 8.4 is used, and clusters 3 and 5 are merged once again. But this time, the average distance for each observation in the merged cluster is used. For example,

$$D\left(C_{1},C_{3,5}\right) = \text{avg}\left(\left(C_{1},C_{3}\right),\left(C_{1},C_{5}\right)\right) = \text{avg}\left(3.16, 3.61\right) = 3.16$$

Table 8.4 Initial distance matrix

Clusters	#1	#2	#3	#4	#5	#6
#1	0.00	1.41	3.16	5.66	3.61	3.16
#2	1.41	0.00	2.00	4.24	2.24	2.83
#3	3.16	2.00	0.00	3.16	**1.00**	4.47
#4	5.66	4.24	3.16	0.00	2.24	5.10
#5	3.61	2.24	**1.00**	2.24	0.00	4.12
#6	3.16	2.83	4.47	5.10	4.12	0.00

Table 8.5 Iteration 2—distance matrix

Clusters	#1	#2	#4	#3 & #5	#6
#1	0.00	**1.41**	5.66	3.16	3.16
#2	**1.41**	0.00	4.24	2.00	2.83
#4	5.66	4.24	0.00	2.24	5.10
#3 & #5	3.16	2.00	2.24	0.00	4.12
#6	3.16	2.83	5.10	4.12	0.00

The resulting matrixes, along with the chart and dendrogram, are given in Tables 8.8, 8.9, and 8.10 and Figs. 8.4 and 8.5.

Let's apply this approach to the previous simple marketing segmentation example. Instead of evaluating the distances between observations, Ward's method uses a minimum variance approach and minimizes the error sum of squares. Like the other methods, each observation starts as its own cluster. E denotes the error sum of squares, which is the total sum of squares for all the clusters. The sum of squares for a given cluster is equal to the squared difference of each observation from its corresponding cluster mean.

Initially, each observation is a cluster, so the mean value of each cluster is its value. For example, observation #1, $x = 8$, so $\bar{x} = 8$ and $y = 5$, so $\bar{y} = 5$. The error sum of squares for observation #1: $(8 - 8)^2 + (5 - 5)^2 = 0$. The total sum of squares $E = 0$. The first step is to determine which two clusters should be combined. Every possible combination will be evaluated, and the combination resulting in the lowest E value will be combined. Let's look at one possible combination, combining clusters #1 and #2. Keep in mind if a cluster consists of only one observation, the error sum of squares is zero. Thus, when clusters 1 and 2 are joined, the total sum of squares for all of the clusters combined is just the error sums of squares for the (1,2) cluster:

$$E_{1,2} = \left(x_1 - \overline{x_{1,2}}\right)^2 + \left(y_1 - \overline{y_{1,2}}\right)^2 + \left(x_2 - \overline{x_{1,2}}\right)^2 + \left(y_2 - \overline{y_{1,2}}\right)^2.$$

$$E_{1,2} = (8 - 7.5)^2 + (5 - 4.5)^2 + (7 - 7.5)^2 + (4 - 4.5)^2 = 1$$

Table 8.6 Iteration 3—distance matrix

Clusters	#1 and #2	#4	#3 and #5	#6
#1 and #2	0.00	4.24	2.00	2.83
#4	4,0.24	0.00	2.24	5.10
#3 and #5	2.00	2.24	0.00	4.12
#6	2.83	5.10	4.12	0.00

Table 8.7 Iteration 4—distance matrix

Clusters	#1, #2, #3, and 5	#4	#6
#1, #2, #3, and #5	0.00	**2.24**	2.83
#4	**2.24**	0.00	5.10
#6	2.83	5.10	0.00

Figure 8.6 shows the resulting E values for each possible clustering combination. The combination of clusters #3 and #5 yields the lowest E value of 0.5 so they are the first two clusters to be joined.

The next step is to create a new table of error sums of squares containing every possible combination. Figure 8.7 shows the resulting E values for the remaining iterations. Let's look at some of the calculations. The E value for the cluster pattern (1,3 5), 2, 4, and 6 is

$$E_{1,3,5} = \left(x_1 - \overline{x_{1,3,5}}\right)^2 + \left(y_1 - \overline{y_{1,3,5}}\right)^2 + \left(x_3 - \overline{x_{1,3,5}}\right)^2 + \left(y_3 - \overline{y_{1,3,5}}\right)^2 + \left(x_5 - \overline{x_{1,3,5}}\right)^2 + \left(y_5 - \overline{y_{1,3,5}}\right)^2$$

$$E_{1,3,5} = (8-6)^2 + (5-4)^2 + (5-6)^2 + (4-4)^2 + (4-6)^2 + (1-4)^2 = 8$$

The E value for the cluster pattern (1,2), (3,5), 4, and 6 is

$$E_{(1,2)(3,5)} = \left(x_1 - \overline{x_{1,2}}\right)^2 + \left(y_1 - \overline{y_{1,2}}\right)^2 + \left(x_2 - \overline{x_{1,2}}\right)^2$$
$$+ \left(y_2 - \overline{y_{1,2}}\right)^2 + \left(x_3 - \overline{x_{3,5}}\right)^2 + \left(y_3 - \overline{y_{3,5}}\right)^2 + \left(x_5 - \overline{x_{3,5}}\right)^2 + \left(y_5 - \overline{y_{3,5}}\right)^2$$

$$E_{(1,2)(3,5)} = E_{(1,2)} + E_{(3,5)} = (8-7.5)^2 + (5-4.5)^2 + (7-7.5)^2$$
$$+ (4-4.5)^2 + (5-5)^2 + (4-3.5)^2 + (5-5)^2 + (3-3.5)^2 = 1+0.5 = 1.5$$

Looking at iteration 2 in the table, the next two clusters that should be combined are #1 and #2 with an E value of 1.5. The grouping of the clusters is shown in Fig. 8.8.

Method Comparison

There has been a significant research on comparing different clustering techniques. There is no one "best" method for cluster analysis. It is dependent on your data and the criteria for your project. For example, Hands and Everitt (1987) compared the five hierarchical clustering techniques mentioned above on multivariate binary data. They found that Ward's method was the best overall. Ferreira and Hitchcock (2009) compared the performance of single linkage, complete linkage, average linkage, and Ward's method for clustering functional data. They too found that Ward's method was usually the best, except when there were large differences among cluster sizes. Average linkage is recommended if the suspected clustering structure has one or two very large groups, particularly when the data is not periodic.

8.3.2 Divisive Clustering

In divisive clustering the clustering is top-down, and all the observations start in one cluster. The clusters are subdivided into smaller and smaller clusters until each observation is its own cluster or until the desired number of clusters is reached. This subdivision is accomplished by finding a single observation that is most dissimilar within a cluster.

Like the agglomerative method, different measurements can be used to determine which observation is most dissimilar. For example, the sums of squared errors or the average distance are two common measurements. Let's apply the divisive method to the data in Fig. 8.2 and use the average distance to determine the

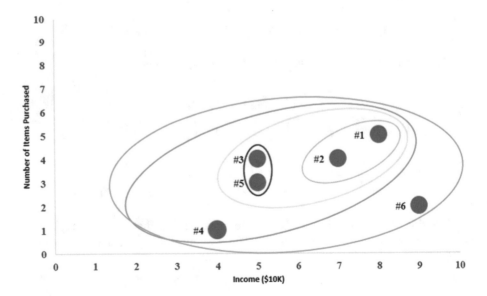

Fig. 8.2 Single linkage grouping

Fig. 8.3 Single linkage dendrogram

Table 8.8 Average linkage iteration 1—distance matrix

Clusters	#1	#2	#4	#3 and #5	#6
#1	0.00	**1.41**	5.66	3.38	3.16
#2	**1.41**	0.00	4.24	2.12	2.83
#4	5.66	4.24	0.00	2.70	5.10
#3 and #5	3.38	2.12	2.70	0.00	4.30
#6	3.16	2.83	5.10	4.30	0.00

Table 8.9 Average linkage iteration 2—distance matrix

Clusters	#1 and #2	#4	#3 and #5	#6
#1 and #2	0.00	4.95	2.75	3.00
#4	4.95	0.00	**2.70**	5.10
#3 and #5	2.75	**2.70**	0.00	4.30
#6	3.00	5.10	4.30	0.00

Table 8.10 Average linkage iteration 3—distance matrix

Clusters	#1 and #2	#3, #4, and #5	#6
#1 and #2	0.00	**3.48**	3.00
#3, #4, and #5	**3.48**	0.00	4.56
#6	3.00	4.56	0.00

observation that is most dissimilar. Keep in mind that at this point all the observations are in one cluster and each observation is not its own cluster. The Euclidean distance matrix is the same values as calculated earlier in Table 8.4.

Step 1. Calculate the average distance of each observation to all other observations.
Average (Dist.) #1 = Avg (1.41, 3.16, 5.66, 3.61, 3.16) = 3.4
Average (Dist.) #2 = 2.54
Average (Dist.) #3 = 2.76
Average (Dist.) #4 = 4.33 <=

Average (Dist.) #5 = 2.64

Average (Dist.) #6 = 3.94

Step 2. Based upon the averages, observation #4 is the most dissimilar and will become its own cluster. Therefore, we have two clusters. Cluster 1, C1, contains observations (#1, #2, #3, #5, #6), and C2 contains observation #4. The next step is to recalculate the distances. The average distance within cluster 1 and to cluster 2 for each observation must be completed, and a decision must be made if that observation is more similar to cluster 1 or cluster 2. Let α = average distance to C1 and β = average distance to C2. If $\alpha > \beta$, then that observation is more similar to cluster 2 and will join C2. Table 8.11 shows the calculations and results.

Observation #5 will join cluster 2. The current clustering consists of C1 (#1, #2, #3, #6) and C2 (#4, #5). The next step is to repeat the same process to see if observations in cluster1 are now more like the revised cluster 2. Table 8.12 shows these results.

Observation #3 now joins cluster 2. Our current clustering consists of C1 (#1, #2, #6) and C2 (#3, #4, #5). This process repeats. The results are shown in Table 8.13.

We have completed the first subdivision of clusters. There are no observations that are more similar to the revised cluster 2. We now have two clusters: C1 (#1, #2, #6) and C2 (#3, #4, #5).

The next step is to split one of the two clusters. The split can be based on the clusters with the largest number of observations (there is a tie in this case), or the split can be based on the cluster with the largest diameter (cluster 1 = 3.16 and cluster 2 = 4.47). The split will be on the observation within cluster 2 that is most dissimilar. It is the observation with the largest average distance to all other observations within cluster 2. The process would repeat to see which observations within cluster 2 are more similar to the new cluster 3. Figure 8.9 shows the completed hierarchy.

Minimal Spanning Tree

Another divisive method that can be used is the minimal spanning tree. Every observation will be connected to every other observation (Fig. 8.10).

Each line will have a weight that is equal to the distance between the two observations. Assume the Euclidean distance measure is used. The minimal spanning tree states that the sum of the weights (distances) which are included in the final spanning tree should be at a minimum. The first step is to determine all of the weight and place them in ascending order. Figure 8.11 shows the ordered table.

The minimum distance is between observations 3 and 5, with a value of 1.00. Let observation 3 be the root of the spanning tree. Connect the other observation by what is called an *edge*, whose weight is the distance between the two observations (Fig. 8.12). The next smallest distance in the list is 1–2. This does not have any common observations with the current tree (3–5); therefore 1–2 will form a second tree whose weight is 1.41. The third distance 2–3 has an observation in common with the first tree (observation 3) and the second tree (observation 2). Observation 3 will connect with observation 2 to join the two trees. The fourth distance on the list is 2–5. Observations 2 and 5 are already in the tree so we skip to the next distance on the list (4–5). Observation 5 is part of the tree. So, we will join observation 4 to

Fig. 8.4 Average linkage grouping

Fig. 8.5 Average linkage dendrogram

observation 5. The next distance on the 2–6 tree will join observation 6 to observation 2. The minimum spanning tree is complete when all the observations have been added. There should never be a closed circuit, within the spanning tree. This indicates that an extra, unnecessary connection was made.

Fig. 8.6 First iteration
using Ward's method

Possible Cluster Combinations					E
(1,2)	3	4	5	6	1
(1,3)	2	4	5	6	5
(1,4)	2	3	5	6	16
(1,5)	2	3	4	6	6.5
(1,6)	2	3	4	5	5
(2,3)	1	4	5	6	2
(2,4)	1	3	5	6	9
(2, 5)	1	3	4	6	2.5
(2, 6)	1	3	4	5	4
(3,4)	1	2	5	6	5
(3,5)	1	2	4	6	**0.5**
(3,6)	1	2	4	5	10
(4, 5)	1	2	3	6	2.5
(4,6)	1	2	3	5	13
(5,6)	1	2	3	4	8.5

8.3.3   Agglomerative Versus Divisive Clustering

Divisive clustering is more efficient if a complete hierarchy is needed, i.e., from the observations as one cluster to the level of individual observations as its own cluster. However, for a complete hierarchy, divisive clustering is more complex as compared to agglomerative clustering, and as a result the agglomerative method is more widely used.

Divisive algorithms tend to be more accurate. Agglomerative clustering evaluates the local clusters or neighboring observations without initially considering the global distribution of the data. Decisions made early in the process cannot be undone, whereas divisive clustering takes into consideration the global distribution of data when making top-level partitioning decisions.

In one study, Roux (2018) used artificial data sets and real-life data sets and found that the divisive algorithm with the silhouette formula, described later in the chapter, performed very well. In addition, he found a clear trend for the divisive algorithms to be superior to the agglomerative counterpart. But one caveat that was noted that no one model could be defined as the "best." It really depended on the data.

Interation #	Possible Cluster Combinations				E
2	(1,2)	(3,5)	4	6	1.5
	(1,3,5)	2	4	6	8
	(1,4)	2	(3,5)	6	16.5
	(1,6)	2	(3,5)	4	5.5
	(2,3,5)	1	4	6	5
	(2,4)	1	(3,5)	6	9.5
	(2, 6)	1	(3,5)	4	4.5
	(3,4,5)	1	2	6	5.3
	(3,5,6)	1	2	4	12.7
	(4,6)	1	2	(3,5)	13.5
3	(1,2,3,5)	4	6		8.8
	(1,2,4)	(3,5)	6		17.8
	(1,2,6)	(3,5)	4		7.2
	(1,2)	(3,4,5)	6		6.3
	(1,2)	(3,5,6)	4		13.7
	(1,2)	(3,5)	(4,6)		14.5
4	(1,2,3,4,5)	6			20
	(1,2,6)	(3,4,5)			12
	(1,2)	(3,4,5,6)			20.8
5	(1,2,3,4,5,6)				30.17

Fig. 8.7 Ward's iteration table

8.4 Non-hierarchical Clustering

In non-hierarchical clustering techniques, a hierarchical structure is not created. This allows for few calculations in which the distances between all pairs of subjects do not have to be calculated. Therefore, this technique may be more efficient when working with very large data sets. Another difference from the hierarchical method is that in non-hierarchical methods the number of clusters needs to be designated in advance. The most common type of non-hierarchical method is k-means clustering.

8.4.1 K-Means Clustering

K-Means clustering is a distance-based technique that calculates the distances between centroids to assign an observation or object to a cluster. The most representative point within the group is called the centroid. Usually, this is the average of the values of the points of data in the cluster. K-Means clustering is one of the simplest

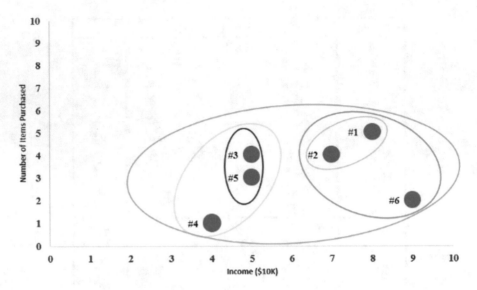

Fig. 8.8 Grouping using Ward's method

Table 8.11 Iteration 1 divisive clustering example

Observation	α	β	Cluster
#1	Avg (1.41,3.16,3.61,3.16) = 2.84	5.6	2.84 < 5.6 Cluster 1
#2	2.12	4.24	Cluster 1
#3	2.66	3.16	Cluster 1
#5	2.74	2.24	Cluster 2
#6	3.65	5.10	Cluster 1

and most popular clustering techniques or algorithms. The K-means cluster identifies k number of centroids and then assigns every observation to the nearest cluster. This minimizes the sum of the distances between the observations and their respective cluster centroid. The steps for the K-means technique are:

1. Set k as the desired number of clusters. The elbow method is one method that can be used to determine the optimal number of clusters. This method is discussed later in the chapter.
2. Select k random points from the data as centroids. These points are often referred to as the seeds, the growing points of the clusters. There are several approaches to select the centroid some of them include:

   (a) Pick k actual observations as the centroids.
   (b) Select random points within the general data space.
   (c) Select points within dense regions.
   (d) Space them uniformly around the data space.
   (e) Furthest point (max-min).

Table 8.12 Iteration 2 divisive clustering example

Observation	α	β	Cluster
#1	Avg (1.41,3.16,3.16) = 2.58	Avg (3.61,5.6) = 4.63	Cluster 1
#2	2.08	3.24	Cluster 1
#3	3.24	2.08	Cluster 2
#6	3.48	4.61	Cluster 1

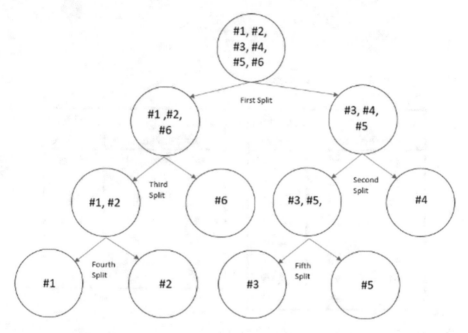

Fig. 8.9 Completed divisive

Table 8.13 Iteration 3 divisive clustering example

Observation	α	β	Cluster
#1	Avg (1.41, 3.16) = 2.29	Avg (3.16, 3.61, 5.6) = 4.14	Cluster 1
#2	2.12	2.83	Cluster 1
#6	3.00	4.56	Cluster 1

3. Assign all the observations to the closest centroid.
4. Then recompute the centroids based upon the newly formed clusters.
5. Repeat steps 3 and 4 to optimize the positions of the centroids. The process is ended when one of the following cases become true: there is no change in their clusters because the clustering has been completed, the results are within a certain threshold level, or the defined number of iterations has been achieved. A common threshold level is to keep iterating until 1% of the observations change their cluster location.

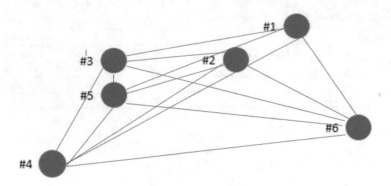

Fig. 8.10 Minimal spanning tree example

Observation		Euclidean Distance		Observation		Euclidean Distance
#3	#5	1.00		#3	#4	3.16
#1	#2	1.41		#1	#5	3.61
#2	#3	2.00		#5	#6	4.12
#2	#5	2.24		#2	#4	4.24
#4	#5	2.24		#3	#6	4.47
#2	#6	2.83		#4	#6	5.10
#1	#3	3.16		#1	#4	5.66
#1	#6	3.16				

Fig. 8.11 Minimal spanning tree ordered distance list

Let's go through the steps with an example. Consider the previous data (Table 8.3).

Step 1—Different from the hierarchical method, an initial number of clusters needs to be selected. Assume for this example that the desired number of clusters is 2 ($k = 2$).

Step 2—For this example we will select two observations 3 and 5, as centroids, indicated in red in Fig. 8.17. Keep in mind that normally these are randomly selected.

Step 3—Using the Euclidean method, as described in Chap. 3, the observations that are closest to each of centroids are assigned to that cluster. The resulting values and clusters are shown in Fig. 8.13.

Step 4—The centroids are recomputed based on the average x and y values of all members in each of the clusters. In Fig. 8.14 the two centroids are denoted as C1 (6.67, 4.33) and C2 (6,2).

Step 5—The clusters are complete. The observations in each cluster remain unchanged. It is important to note that if two different centroids were selected the

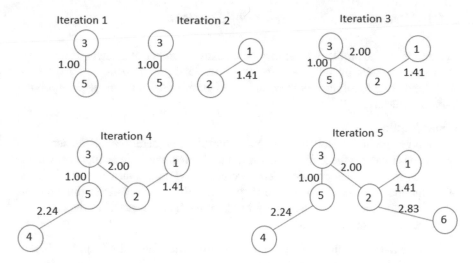

Fig. 8.12 Minimal spanning tree iterations

clustering results would change. For example, if you were to start with observations 2 and 6 as the centroids, the resulting clusters would differ. As seen in Fig. 8.15, customer #6 is now its own cluster.

8.4.2 Initial Centroid Selection

The number of iterations needed to converge on an optimal set of clusters is dependent on the initial centroid selection. One variant of K-means clustering called K-means++ can be used to optimize the initial centroid at the start of K-means clustering process. The steps for the K-means++ are:

1. Randomly select the first centroid.
2. Calculate the distance of each observation from the centroid.
3. Choose the next centroid that has the largest squared distance from the centroid.
4. Repeat process 2 and 3 using the distances to the closest centroid.

Applying the algorithm from the previous example, observation 3 is randomly selected as the initial observation. The squared distances from each observation to observation 3 are then calculated. The observation with the largest squared distance from the centroid is then selected as the next centroid. The calculations and results are shown in Figs. 8.16 and 8.17. The optimal starting centroids are observations 3 and 6.

8.4.3 Determining the Number of Clusters

Depending on the problem, there may be a desired number of clusters. If not, there are many methods including elbow, silhouette, and gap statistic that have been developed for identifying the recommended number of clusters when applying *K*-means clustering.

Elbow Method

The elbow method is the most popular method for determining clusters. The approach is to repeat the *K*-means algorithm multiple times with an increasing number of clusters and then plot a clustering score, based on the variation as a function of the number of clusters. The method is based on the theory of diminishing returns where an additional cluster does not provide a much better model of the data. The steps for the elbow method are:

1. Use the *K*-means clustering technique for different values of *k* (e.g., 1 to 10).
2. For each *k*, calculate the variation score.
3. Plot the variation score with the number of clusters *k*. The variation score that is most used in the elbow method is the *distortion score*, which is the sum of squared distances from each observation to its corresponding centroid.
4. The location of the bend (elbow) in the plot is the recommended number of clusters to use.

For illustration purposes, if the results of running the elbow method yielded the plot shown in Fig. 8.18, the optimal number of clusters would be 3, the point where the line flattens out.

Average Silhouette Method

The average silhouette method first described by Kaufman and Rousseeuw (1990) is often used for validating clusters of data and can also be used to determine *k*. It measures how similar an observation is to its assigned cluster (*cohesion*) compared to other clusters (*separation*). It determines how well each observation lies within its cluster. The silhouette coefficient values range from −1 to 1. The higher the average silhouette score, the better the clustering. Silhouette coefficients near positive 1 indicate that the observation is far away from the neighboring clusters. A value of 0 indicates that the observation is near or in between the two neighboring clusters and it is indifferent between the two clusters. A negative value indicates that the observation may have been assigned to the wrong cluster. The *k* that maximizes the average silhouette is the optimal number of clusters. The formula to calculate the silhouette score for each data value, *i*, is

$$s_i = \frac{b_i - a_i}{\max\left(bi, ai\right)}$$

where a_i measures the distance between the observation and all other observations within the cluster and b_i measures the separation, the distance between the observations, and the next closest cluster center.

| Observations | | Euclidean |
Start	End	Distance
#1	#3	3.16
#1	#5	3.61
#2	#3	2.00
#2	#5	2.24
#4	#3	3.16
#4	#5	2.24
#6	#3	4.47
#6	#5	4.12

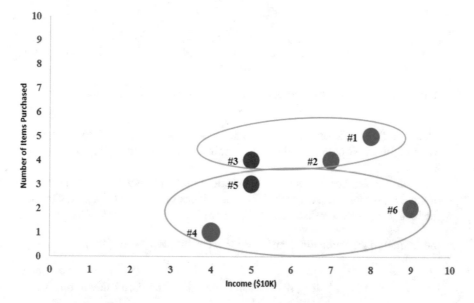

Fig. 8.13 Customer income and number of observations—initial clusters

Gap Statistic

The gap statistic was first developed by Tibshirani et al. (2001). It compares the total within-cluster variation for different values of k with their expected value under the null reference distribution of the data samples (i.e., a distribution with no obvious clustering—this indicates a random uniform distribution). This method works well at identifying the number of clusters when the clusters are well separated.

The steps for the gap statistic are:

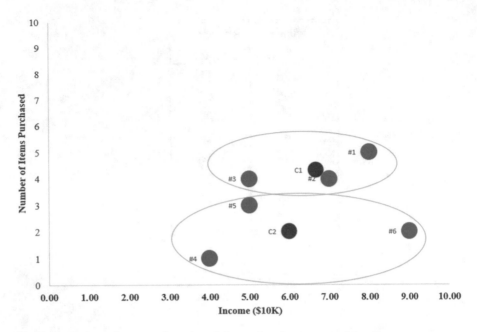

Fig. 8.14 Customer income and number of observation cluster—second iteration

1. Cluster the observations, varying the number of clusters from 1 to some maximum number, K. Compute the total within-cluster distances W_k, where $k = 1, 2, \ldots, K$.
2. Generate β reference data sets with a random uniform distribution. Cluster each of these reference data sets with a varying number of clusters $k = 1 \ldots K$. Compute the total within-cluster distances W_{kb}^*, where $b = 1 \ldots \beta$, $k = 1 \ldots K$. Then calculate the estimated gap statistic: $\mathrm{Gap}(k) = \left(\frac{1}{\beta} \right) \sum_b \log\left(W_{kb}^* \right) - \log\left(W_k \right)$.

3. Let $\bar{l} = \left(\frac{1}{\beta} \right) \sum_b \log\left(W_{kb}^* \right)$, calculate the standard deviation $\mathrm{sd}_k =$

   $[\left(\frac{1}{\beta} \right) \sum_b \{\log\left(W_{kb}^* \right) - \bar{l} \}^2]^{1/2}$ and let $s_k = \mathrm{sd}_k \sqrt{\left(1 + \frac{1}{\beta} \right)}$.

4. Choose the cluster number = smallest k such that the gap statistic is within one standard deviation of the gap at $k + 1$: $\mathrm{Gap}(k) \geq \mathrm{Gap}(k + 1) - S_{k+1}$.

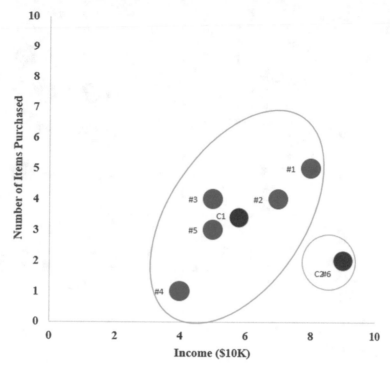

Fig. 8.15 Customer income and number of observations cluster with different starting centroids

Fig. 8.16 Initial centroids using *K*-means++

Observations		Distance	Distance^2
Start	End		
#3	#1	3.16	10
#3	#2	2.00	4
#3	#4	3.16	10
#3	#5	1.00	1
#3	#6	3.61	13

8.4.4 Evaluating Your Clusters

Keep in mind that there is not always a "optimal" number of clusters. It is dependent on the data and the goals of the project. Experts in the data domain can be used to assess the validity of the clusters. Alternatively, there are several numerical metrics that can be used. These statistics can be classified into two classes: internal and external validation metrics. Internal validation is used to measure the goodness of clustering without knowledge of external information. External validation is used to measure the extent to which cluster labels match externally supplied class labels.

Fig. 8.17 Graph of initial centroids using *K*-means++

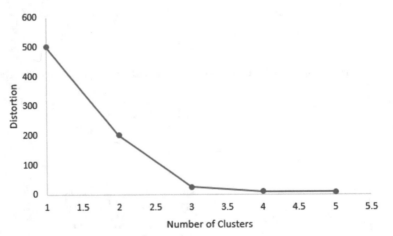

Fig. 8.18 Distortion vs number of clusters

Internal measures are the most widely used clustering validation metrics. Two such metrics include the silhouette and Dunn's index. The silhouette index was described earlier in this chapter to determine the number of clusters. It is based on the pairwise difference of between and within-cluster distances. Dunn's index (DI) measures the compactness and separation of clusters. A higher DI value implies better clustering, assuming that the goal is to have compact and well-separated clusters. Dunn's index is equal to the minimum distance between observations not in the same cluster divided by the maximum cluster size (largest within-cluster distance). Well-separated and smaller, more compact clusters lead to higher DI values.

External measures are frequently used when new clustering techniques are developed, and the goal is to compare it to existing techniques. The Rand index and Jaccard coefficient are similarity-oriented measures. They measure the similarity between two data clusters. A Rand index value of 0 indicates that the observations were assigned into clusters randomly. A value of 1 indicates that the two clusters are identical. The Jaccard coefficient is also a measure of similarity for two sets of data, with a range from 0% to 100%. The higher the percentage, the more similar the two clusters.

When viewing the results of clustering, there are other statistics that may be examined. They include the R^2, the cubic clustering criterion (CCC), and pseudo-F. The R^2 measures the extent to which the clusters are different from one another. An R^2 value closer to 1 is desirable.

The CCC compares the R^2 value from a set of clusters, with the R^2 value calculated by clustering a uniformly distributed set of data. Large positive values of the CCC indicate a better clustering. However, the CCC may be incorrect if clustering variables are highly correlated.

The pseudo-F statistic is used to explain the compactness of clusters and can also be used to determine the number of clusters. It compares the between-cluster to the within-cluster sum of squares. A large pseudo-F statistic indicates better clustering. As the pseudo-F value increases with k, the value of k near the maximum pseudo-F is a good choice for the number of clusters.

8.5   Hierarchical Versus Non-hierarchical

Hierarchical clustering doesn't handle large data sets as well as K-means clustering. The time complexity of K-means clustering is linear while hierarchical clustering is quadratic. In hierarchical clustering the results are consistent and reproducible, whereas, with K-means clustering, we start with a random set of clusters, so the results are generally not reproducible. In addition, K-means clustering requires the pre-selection of the number of clusters.

8.6   Cluster Analysis Using SAS Enterprise Miner™

In SAS Enterprise Miner™, there are two nodes that can be used to cluster data. The cluster node and the SOM/Kohonen node (Fig. 8.19). The SOM/Kohonen node performs cluster analysis through network training. Kohonen SOMs are similar to a neural network design where each input unit is connected to each output unit. For most cluster analysis situations, the clustering node is recommended over the SOM/Kohonen node.

Fig. 8.19 Cluster and
SOM/Kohonen nodes

8.6.1   Cluster Node

The cluster node has a specification method that is used to determine the maximum
number of clusters. The two options are automatic or user-specified. If set to automatic, SAS Enterprise Miner™ will determine the optimum number of clusters to
create. When set to automatic, the value in the maximum number of clusters is not
used. If it is set to user-specified, the maximum value in the selection criterion properties can be set. Let's first review how the cluster node works.

Step 1. The default of 50 cluster seeds is first chosen. The default value is used
when the preliminary maximum value in the selection criterion properties is not set.
The observations in the training data are then assigned to the closest cluster seed.
This creates an initial clustering or preliminary pass. Next, the means of each variable in each of these preliminary clusters are substituted for the original clusters.

Step 2. Ward's clustering method (default) is used to merge the clusters that were
formed in the first step. Alternatively, the clustering method set in the specification
method property can be used to change the method to average or centroid. Then, the
smallest number of clusters that meets the following three criteria is selected:

- The number of clusters must be greater than or equal to the number specified in
  the minimum value of the selection criterion properties and less than or equal to
  the final maximum value.
- The number of clusters must have a CCC value greater than the CCC cutoff value
  set in the selection criterion properties.
- A peak in the number of clusters exists.

If the data cannot be clustered in a manner that meets all three criteria, then the
number of clusters is set to the first local peak.

Step 3.   The number of clusters determined by the second step provides
the value for k in a K-means clustering of the original training data cases.

8.6.2   Additional Key Properties of the Cluster Node

Other key properties of the cluster node include:

Internal Standardization Property The Internal Standardization property is used
when creating the clusters. The three options are none, standardization (default),
and range. Standardization helps make the relative weight of each variable equal by

converting each variable to a unitless measure or relative distance. By standardizing, the numerical variables will have a mean of zero and standard deviation of one (converted to z-scores).

Encoding of Class Variables The encoding method specifies how to compute one or more quantitative variables to be used for computing distances. These quantitative variables are like the dummy variables that are used in linear models. The number of quantitative variables resulting from the encoding of the class variables is called the dimension.

Ordinal Encoding—Ordinal encoding is used to specify the encoding method for
   ordinal class variables. The options are rank (default), index, bathtub, and thermometer. The rank and index encoding methods have a dimension of 1, whereas
   the thermometer and bathtub encoding methods have a dimension of the number
   of categories minus 1.

Nominal Encoding—Normal encoding is used to specify the encoding method for
   nominal class variables. The options are GLM (default), deviation, and reference. The GLM encoding method has a dimension equal to the number of categories, whereas the deviation and reference encoding methods have a dimension
   equal to the number of categories minus 1.

Initial Cluster Seeds Initial cluster seeds are used in step 1 of the process. Initial cluster seeds specify how the first cluster seeds are created. The first initial seed is usually the first complete (no missing values) observation. Then the nearest complete observation that is within the distance specified in the Minimum Radius property is used as the second seed.

Seed Initialization Method If the seed initialization method is set to default and the Drift During Training property is set to Yes, then MacQueen's K-means algorithm is used to compute the initial seeds. Otherwise, partial replacement is the default. If the drift option is set to Yes, temporary clusters are created by assigning each observation to the nearest seed. Once an observation has been assigned to a cluster, that cluster's seed is recalculated to the mean of the observations currently assigned to the cluster. Thus, the cluster seeds are considered to "drift" rather than remaining constant.

Missing Values Missing values allows you to specify how to impute any missing observations (values).

Fig. 8.20 Cluster analysis variables

8.7   Applying Cluster Analysis to the Insurance Claim Fraud Data Set

Cluster analysis can be applied to the automobile insurance claim fraud train data set. To keep the clusters simple, let's limit the number of variables to five, Employment_Status, Fraudulent_Claim (target), Gender, Marital_Status, and Vehicle_Class (Fig. 8.20)

Create and connect a cluster node to the data set. The cluster node is located on the Explore tab. Set the properties of the cluster node as shown in Fig. 8.21. If your numerical variables are not measured in the same units (e.g., dollars), then they should be standardized. The specification method for the number of clusters is set to automatic. Ward's, average, and centroid methods for finding the number of clusters can be used. After the number of clusters is determined, the clusters are then calculated using the K-means algorithm. To keep the problem relatively simple, limit the number of clusters to a maximum of 10, and use the centroid clustering method. The other two clustering methods available are average linkage and Ward's. Run and view the results.

The results indicate the best solution is one with four clusters. The segment size chart (Fig. 8.22) shows the relative size of the different clusters or segments. If you hover your mouse over the various segments, SAS will display the number of

Fig. 8.21 Cluster node
properties

Train	
Variables	...
Internal Standardization	Standardization
⊟ Number of Clusters	
⋮ Specification Method	Automatic
⋯ Maximum Number of Clusters	10
⊟ Selection Criterion	
⋮ Clustering Method	Ward
⋮ Preliminary Maximum	10
⋮ Minimum	2
⋮ Final Maximum	10
⋯ CCC Cutoff	3
⊟ Encoding of Class Variables	
⋮ Ordinal Encoding	Rank
⋯ Nominal Encoding	GLM
⊟ Initial Cluster Seeds	
⋮ Seed Initialization Method	Default
⋮ Minimum Radius	0.0
⋯ Drift During Training	No
⊟ Training Options	
⋮ Use Defaults	Yes
⋯ Settings	...
⊟ Missing Values	
⋮ Interval Variables	Default
⋮ Nominal Variables	Default
⋮ Ordinal Variables	Default
⋯ Scoring Imputation Method	None

Fig. 8.22 Segment size

Fig. 8.23 Segment plot

Segment Id	Frequency of Cluster	Root-Mean-Square Standard Deviation	Maximum Distance from Cluster Seed	Nearest Cluster	Distance to Nearest Cluster	Employment _Status=Dis abled	Employment _Status=Em ployed	Employment _Status=Me dical Leave	Employment _Status=Reti red	Employment _Status=Une mployed
1	745	0.451569	3.572764	3	2.115492	0.07651	0.651007	0.04698	0.084564	0.14094
2	78	0.396607	2.639303	4	3.435535	0.038462	0.679487	0.012821	2.78E-17	0.269231
3	2061	0.386848	3.792256	4	2.001327	0.040272	0.612809	0.049005	0.019408	0.278506
4	2117	0.391708	3.559969	3	2.001327	0.033066	0.625886	0.04582	0.02598	0.269249

Fig. 8.24 Mean statistics table—Employment_Status columns

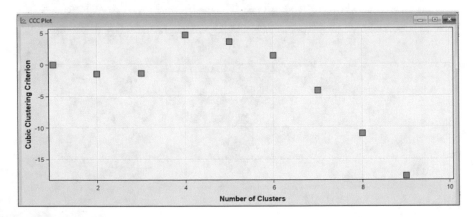

Fig. 8.25 CCC plot

```
23      The CLUSTER Procedure
24      Ward's Minimum Variance Cluster Analysis
25
26                   Eigenvalues of the Covariance Matrix
27
28              Eigenvalue     Difference      Proportion      Cumulative
29
30       1      0.47503725     0.29155361       0.5136          0.5136
31       2      0.18348364     0.01494581       0.1984          0.7119
32       3      0.16853783     0.11980537       0.1822          0.8941
33       4      0.04873246     0.01791267       0.0527          0.9468
34       5      0.03081980     0.01313535       0.0333          0.9801
35       6      0.01768445     0.01728135       0.0191          0.9992
36       7      0.00040310     0.00010507       0.0004          0.9997
37       8      0.00029803     0.00028970       0.0003          1.0000
38       9      0.00000833     0.00000833       0.0000          1.0000
39      10      0.00000000     0.00000000       0.0000          1.0000
40      11      0.00000000     0.00000000       0.0000          1.0000
41      12      0.00000000     0.00000000       0.0000          1.0000
42      13      0.00000000     0.00000000       0.0000          1.0000
43      14     -.00000000      0.00000000      -0.0000          1.0000
44      15     -.00000000      0.00000000      -0.0000          1.0000
45      16     -.00000000                      -0.0000          1.0000
46
47     Root-Mean-Square Total-Sample Standard Deviation      0.433189
48
49     Root-Mean-Square Distance Between Observations       2.450485
50
```

Fig. 8.26 Output Window—eigenvalues

observations within that cluster. The chart indicates there are 2117 observations that are in the largest cluster, cluster 4.

The characteristics of each of the clusters can be determined by looking at the segment plot (Fig. 8.23) and the mean statistics table (Fig. 8.24). The employment status factor plot visually shows the distribution of employment across the clusters. The exact values are given in the mean statistics table. Looking at the employed column, cluster 1 has 65% employed, cluster 2 has 68%, cluster 3 has 61%, and cluster 4 has 63% employed. The gender plot shows that cluster 3 is 100% female and cluster 4 is 100% male. The marital status chart shows that 100% of cluster 1 is divorced, and the vehicle class chart indicates that cluster 2 consists of 100% luxury car owners.

The cubic clustering criterion (CCC) plot can be viewed by selecting the View menu, summary statistics, and CCC plot. The CCC plot (Fig. 8.25) indicates that given the initial setting of a maximum of ten clusters, which was set in the cluster

```
52                                        Cluster History
53    Number
54      of                                        Semipartial                  Pseudo F    Pseudo
55    Clusters    ------Clusters Joined------   Freq   R-Square   R-Square    Statistic   t-Squared   Tie
56
57       9     0B3        0B5               115    0.0005      .308         277         2.8
58       8     0B2        0B4                56    0.0019      .306         314        29.4
59       7     CL8        0B7                78    0.0029      .303         361        32.8
60       6     CL7        CL9               193    0.0064      .296         421        40.2
61       5     0B1        CL6               348    0.0141      .282         491        71.9
62       4     CL5        0B6              1458    0.0208      .262         590         111
63       3     CL4        0B10             2681    0.0543      .207         653         334
64       2     0B8        0B9              2320    0.0585      .149         873         448
65       1     CL3        CL2              5001    0.1487      .000          .          873
66
67
68
69    Candidates for Optimum Number of Clusters
70
71              Number    Clustering
72                of        Cubic
73    Obs     Clusters    Criterion
74
75     1        4        4.68618
76
77
78    *-------------------------------------------------------------*
79    * Score Output
80    *-------------------------------------------------------------*
81
82
83    *-------------------------------------------------------------*
84    * Report Output
85    *-------------------------------------------------------------*
86
87
88
89
90    Variable Importance
91
92                               Number of    Number of
93                               Splitting    Surrogate
94    Variable Name      Label      Rules        Rules      Importance
95
96    Gender             Gender        1            0        1.00000
97    Marital_Status     Marital_Status 1           1        0.98289
98    Vehicle_Class      Vehicle_Class  1           1        0.75355
99    Employment_Status  Employment_Status 0        1        0.71535
```

Fig. 8.27 Output Window—cluster history

Fig. 8.28 Segment Profile node

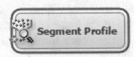

node property window, four clusters are the optimal number of clusters. It is the first and most distinct peak.

More detailed information on the clustering can be found in the output results window. The output (Fig. 8.26) first displays the table of eigenvalues of the covariance matrix, which are used in calculating the cubic clustering criterion. The first column displays each eigenvalue. The second column is the difference between the eigenvalue and its successor. The remaining two columns display the individual and cumulative proportion of variation associated with each eigenvalue. For example, the first two variables account for approximately 71% of the total variation.

Fig. 8.29 Variable worth charts

Fig. 8.30 Segment Profile
node properties

.. Property	Value
☰General	
Number of Midpoints	8
Profile All	No
Cutoff Percentage	95
☰Input Variables	
Number of Inputs	10
Minimum Worth	0.01
Maximum Depth	1
Print Worth Statistics	Yes

Below the eigenvalues table, the output provides the square root of the average of squared errors; in this example it is 0.433. The lower the standard deviation, the more homogeneous the clusters. This statistic is followed by the root-mean-square distance between observations, 2.45. The smaller the value, the more compact the clusters are.

The next table provided is the cluster history (Fig. 8.27). Starting from the top, the first cluster to be created, cluster 9 (CL9), consisted of observations 3 and 5 identified as OB3 and OB5 with 115 observations within the cluster. Then the second cluster (CL8) was created by joining OB2 and OB4 with 56 observations within the cluster. The third cluster (CL7) was created by joining the previously created cluster 8 (CL8) and OB7 with 76 observations.

Fig. 8.31 Profile Segment window

To gain more information about the clusters, a Segment Profile node (Fig. 8.28) located under the Assess tab can be added. The variable worth charts (Fig. 8.29) show each variable worth or contribution to differentiate that cluster from the rest of the sample. The charts are ordered based on the size of the cluster. The first chart shown in the worth window is cluster 4, the largest cluster. The largest differentiator for cluster 4 is gender followed by marital status. Gender is by far the greatest differentiator for clusters 4 and 3. Marital status is the differentiator for cluster 1, and vehicle class is the differentiator for cluster 2. Notice not all the variables are given as having any worth. This is because, in the initial properties of the Segment Profile node (Fig. 8.30), the default minimum worth is set to 0.01. Only those variables with a variable worth of at least 0.01 are shown. This value can be changed if needed.

The Profile Segment window (Fig. 8.31) provides more information about each cluster relative to the sample data. The outer pie shows the distribution of the cluster, and the inner pie shows the distribution of the sample. Cluster 4 is 100% male, whereas the sample population is approximately 50/50. The marital status of cluster 4 is mostly married (64%) with some identifying as single (32%), whereas the sample is married (67%), single (26%), and divorced (15%) observations. In summary, cluster 1 can be classified as the divorcees, cluster 2 as the luxury car owners, cluster 3 as married and single females, and cluster 4 as married and single males.

If the cluster node was run again using the same settings except changing the initial clustering method to centroid. The result would be five clusters, with cluster 1 classified as the luxury car owners, cluster 2 as married and employed females, cluster 3 as married males, cluster 4 as divorcees, and cluster 5 as single and unemployed. In the centroid case, the employment status variable does have a worth value >0.10 in some clusters (Fig. 8.32).

Fig. 8.32 Centroid method—variable worth chart

8.8   Chapter Summary

Cluster analysis is an unsupervised machine learning technique that divides observations or data points into groups that are more similar to one another and dissimilar to other points in other groups. Cluster analysis can be applied in a wide area of fields including science and business.

Two types of clustering methods include hierarchical and non-hierarchical. In hierarchical clustering, the clusters are formed in a tree-like structure or dendrogram, creating a hierarchy (ordering) of clusters. There are two types of hierarchical clustering, agglomerative and divisive. Agglomerative clustering is the most widely used hierarchical clustering method. It is a bottom-up approach where each observation is its own cluster and clusters are merged as you move up the hierarchy. Divisive clustering is a top-down approach. All the observations start as one cluster, and the clusters are divided as you move down the hierarchy.

Non-hierarchical clustering does not form a hierarchical structure. The K-means clustering technique is the common method of non-hierarchical clustering. The basic technique involves first identifying the desired k clusters. Each observation is assigned to one of the k clusters. The method attempts to minimize the distance between the centroid and each observation. There are several methods that can be used to determine k. Some of the methods include the elbow, silhouette, and gap statistic.

In SAS Enterprise Miner™, the Cluster and SOM/Kohonen nodes can be used for cluster analysis. However, the cluster node is recommended for most data analysis cases.

Discussion Questions
1. Discuss how clustering can be used as a method to evaluate a large volume of data.
2. Describe the difference between hierarchical and non-hierarchical clustering.
3. Describe and compare the two types of hierarchical clustering methods.
4. Describe the different methods that can be used to determine the number of clusters.

References

Ferreira L, Hitchcock DB (2009) A comparison of hierarchical methods for clustering functional data. Commun Stat – Simul Comput 38:1925–1949

Hands S, Everitt B (1987) A Monte Carlo study of the recovery of cluster structure in binary data by hierarchical clustering techniques. Multivar Behav Res 22:235–243

Kaufman L, Rousseeuw P (1990) Finding groups in data: an introduction to cluster analysis. Wiley Publishing

Mahalanobis PC (1936) On the generalized distance in statistics. J Genet 41:159–193

Prabhakaran S (2019) Mahalanobis distance – understanding the math with examples (python). Retrieved from https://www.machinelearningplus.com/statistics/mahalanobis-distance/

Roux M (2018) A comparative study of divisive and agglomerative hierarchical clustering algorithms. J Classif, Springer 5(2):345–366

Sokal RR, Michener CD (1958) A statistical method for evaluating systematic relationships. Univ Kansas Sci Bull 28:1409–1438

Tibshirani R, Walther G, Hastie T (2001) Estimating the number of clusters in a dataset via the gap statistic. J R Stat Soc 63(2):411–423

Chapter 9
Text Analytics: Using Qualitative Data to Support Quantitative Results

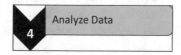

Chapter Learning Objectives
1. Describe text analytics.
2. Differentiate between qualitative and quantitative data.
3. Create a text analytics model.
4. Analyze unstructured text data.

In Chap. 1, we introduced the concept of big data. Recall that in particular the explosive growth in big data has been largely fueled by the massive volumes of unstructured data. Much of this consists of text data. Let's consider one example, Twitter. Twitter began operations in mid-2006. Although it took several years to grow into one of the significant social media forces, it is now estimated that they produce over 12 terabytes of data daily (Matthews 2018). That results in over 4.3 petabytes of data per year. It is one example of a very large source of unstructured data that contains an enormous amount of information about subjects such as product reviews, perceptions of organizations, issues of importance to consumers, and new product or service introductions.

Now let's consider another example, patient medical records. In this case we have a file of a combination of structured and unstructured data consisting of the patient's personal medical information supplemented with detailed notes from the physician that describe the patient and their various medical conditions. Both examples are very different in the size and scope of the data that they deal with. However, in both cases they provide examples of qualitative (i.e., text) data that can be useful to support predictive analytics analysis.

Many organizations have an extensive volume of text information from both internal and external sources. These include documents such as emails, reports, working papers, results of analysis and research, and survey feedback. Technologies such as Microsoft Azure, Amazon AWS, or Hadoop have made it possible to have this information in a readily accessible format that can be accessed by many analysts simultaneously.

9.1   What Is Text Analytics?

Text analytics retrieves and evaluates unstructured data for the purpose of searching for meanings or associations. This form of analysis can be conducted independent of or in conjunction with predictive analytics. When used with predictive analytics, text analytics can help support identification of correlations or associations between variables by evaluating the context and meaning of descriptive information about those variables. Whereas predictive analytics uses mathematical algorithms to evaluate structured data, text analytics primarily uses natural language processing or word extraction to evaluate unstructured data.

Text analytics and text mining are synonymous terms. Text mining provides structure to unstructured data by filtering and extracting only those terms relevant to the subject of analysis. It generally involves a large amount of data, often in the form of documents. These repositories of text data are referred to as a *corpus*. The corpus of documents used in text mining can be composed of many different file formats. For example, they may include:

- .txt—Text file format
- .rtf—Rich Text Format
- .docx—Microsoft Word format
- .pdf—Portable document format
- .jpeg—Joint Photographic Experts Group
- .png—Portable Network Graphics
- .gif—Graphics Interchange Format.

Often, document repositories are composed of many differently formatted documents, and the size of each file may vary greatly.

Chakraborty et al. (2013) define two broad categories of text mining:

1. Search—This includes information retrieval and content categorization.
2. Descriptive and predictive analysis—This includes the actions to analyze text information.

SAS Enterprise Miner™ provides support for text mining through seven nodes that enable specific functionality to process multiple documents to seek specific terms or patterns within the text data.

9.2   Information Retrieval

With all forms of analytics, data is needed. This is also necessary with text mining. *Information retrieval* is a general term to describe the collection of a large amount of text data for the purpose of subsequent analysis. One potential source of these vast amounts of text data is the Internet. *Web crawling* is a term used to define a process for copying and indexing data from websites. Web crawling is commonly used by search engines, such as Google, Bing, or Yahoo, to learn about the content of web pages using *web crawlers*. Web crawlers are also referred to as spiders, spider bots, or web bots. Web crawling is a process that quickly enables the creation of text data for analytics. Subsequent to web crawling is *web scraping*, the extraction of specific data, from specified URLs for further analysis.

In SAS Enterprise Miner™, web crawling can be performed with the use of the text import node (Fig. 9.1).

The text import node is a dual-purpose node. In addition to web crawling, it also performs the importing of a corpus of documents contained within a project library. To perform web crawling, the URL of the website must be included in the node properties (Fig. 9.2). In addition, there must be an Import File Directory and Destination Directory specified. The Import File Directory specifies the location of the web pages to be processed after they are downloaded from the web crawling process. The Destination Directory contains the location for the output files that contain the corpus of documentation created as a result of the web crawling process.

Let's examine the key properties of the text import node.

Train Properties
- *Import File Directory*—The Import File Directory specifies the location of the HTML files that are downloaded from a web scrapping process after execution or the location of a corpus of documents to be imported into a text analytics model.
- *Destination Directory*—The Destination Directory specifies the location of the text (i.e., .txt) files that are created from the text import.
- *Language*—Language specifies the written language(s) of the imported files.
- *Extensions*—Specifies the file type(s) to be imported.
- *Text Size*—Specifies the maximum size of the number of characters that will be created in the output data sets. The maximum size is 32,767.

Web Crawl Properties
- *URL*—This is the home page URL of the web page to be evaluated.

Fig. 9.1 Text Import

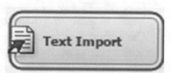

Fig. 9.2 Text Import Train
Properties

.. Property	Value
General	
Node ID	TextImport2
Imported Data	...
Exported Data	...
Notes	...
Train	
Import File Directory	K:\McCarthy\WEBSOUR(...
Destination Directory	K:\McCarthy\WEBDESTI ...
Language	English ...
Extensions	
Text Size	1000
⊟Web Crawl	
⋮URL	http://www.insurance.ca.g(
⋮Depth	1
⋮Domain	Restricted
⋮User Name	
⋮Password	
Status	
Create Time	3/1/19 11:10 AM
Run ID	5006e8a6-f0a7-472b-8806
Last Error	
Last Status	Complete
Last Run Time	2/11/21 3:28 PM
Run Duration	0 Hr. 2 Min. 23.20 Sec.

- *Depth*—Depth specifies the number of levels within the home page URL to be evaluated.
- *Domain*—The domain contains the value restricted or unrestricted which indicates whether the crawl will remain within the specified domain.
- *User Name*—For a secure website, the user name is the login name for the site.
- *Password*—For a secure website, the password for the site must be included.

Execute the node to complete a web crawl. Once the web crawl has completed, the Destination Directory files can serve as the basis for text analysis of web-based content. Alternatively, the text import can also be used to input document files from a variety of formats. Both web content and document files can be combined for analysis.

The automobile claim fraud case includes a corpus of 100 documents that contain notes from claim adjusters during the investigation and settlement of a sample of these automobile insurance claims. A text import of these documents permits all of the documents to be analyzed simultaneously. This can be useful when looking for trends or analyzing patterns.

The output results of the text import provide summary-level information about the entire corpus of documents. Figure 9.3 reflects the proportion of document languages, and Fig. 9.4 reflects the proportion of document types within the corpus.

Note that the maximum text size to be imported for each document is 32,767 bytes; however, the default is 100 bytes. The document length by frequency (Fig. 9.5) can be used to determine if any documents exceed the maximum length (for which truncation will occur) or if the default length setting is too low.

9.3   Text Parsing

Parsing is a term used to describe the process of analyzing sentences into their logical syntactic components. It breaks apart sentences into its components for the purpose of being able to analyze the words that define structure and meaning. It also enables us to separate those words that provide context to the meaning. When parsing text, the *bag of words* method refers to considering every individual word without regard to its context or meaning. Each word is a *token*. Tokens need to be separated into actual words versus those characters that act as separators or numbers and are not meaningful for analysis. *Linguistics* looks beyond the representation of the word to also consider context and meaning. It is in this step that we need to consider issues such as synonyms. Synonyms could be evaluated as individual terms, or they could be grouped together to be considered as the same term. Text parsing begins by identifying all of the tokens within all of the corpus of documents. It will then group tokens into terms. Once we have identified terms, we can then begin to look at context and meaning. To do so, it is useful to identify terms by type, such as nouns, verbs, and adjectives. This can also be referred to as *part of speech* tagging (Abbott 2014). This can potentially result in hundreds or thousands of words, depending upon the size of the corpus. Further it is important to remember that some words may have multiple meanings or even be ambiguous.

The text parsing node in SAS Enterprise Miner™ (Fig. 9.6) is used to begin text analysis by performing the processes described above. We begin by identifying the language that will be utilized.

Subsequently, it enables *ignores* to be defined. Ignores are parts of speech that are not relevant for text analysis. Parsing enables a filter to be applied through the use of a *start list* and *stop list* (Fig. 9.7).

The start list enables a dictionary of terms that will be used for analysis. It is a representation of the terms that are unique to the subject of analysis. For example,

Fig. 9.3 Document languages

Fig. 9.4 Document types

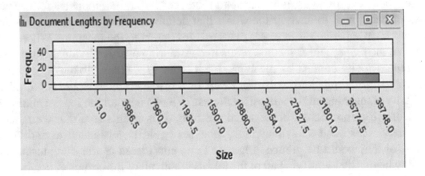

Fig. 9.5 Document length by frequency

if the text analysis deals with medical documents, a start list could be used to define specific medical terms that are the focus of the analysis. A stop list enables terms to be immediately eliminated from further analysis.

Text parsing supports *stemming*. Stemming is the ability to combine similar words, such as verb tenses or singular and plural nouns. For example, the words run, ran, and running would all be considered three different tokens. However, stemming combines them into a single term for analysis. Though stemming could be considered as a rules-based process, it becomes more complex when we consider that plural nouns are not as simple as adding the letter s in all cases. Text parsing helps to reduce *noise* in text data. Noise in text data are words that are not relevant for descriptive or predictive analysis. Parsing can condense a large volume of tokens into a more manageable subset.

Let's examine the key properties of the text parsing node.

Train Properties
- *Language*—Language specifies the vocabulary that will be used for parsing.
- *Different Parts of Speech*—Different Parts of Speech is a Yes/No indicator that identifies if the output tokens will be grouped by their role (e.g., noun, verb, adjective).
- *Noun Groups*—Noun Groups are a Yes/No indicator that identifies if similar nouns will be grouped together.

Fig. 9.6 Text parsing

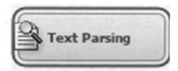

Fig. 9.7 Text parsing properties

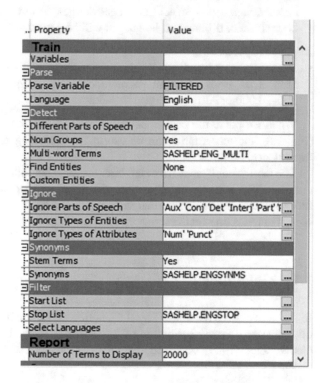

- *Multi-word Terms*—Multi-word Terms specify a data set that contains multi-word terms; a default file of terms is provided.
- *Find Entities*—Find Entities specifies whether entities will be identified. Named entities are a natural language parsing technique to classify nouns, such as identifying them as a person or place (i.e., location). Acceptable values are None, Custom, Standard, or All.
- *Custom Entities*—Custom Entities are used to define the specific entities when the Find Entities property is set to Custom.
- *Ignore Parts of Speech*—Ignore Parts of Speech specifies parts of speech to be filtered.
- *Ignore Types of Entities*—Ignore Types of Entities specifies the types of entities to be filtered.
- *Ignore Types of Attributes*—Ignore Types of Attributes specifies the attributes to be filtered. An attribute indicates the character types that are part of a term.
- *Stem Terms*—Stem Terms is a Yes/No indicator to turn stemming on/off.

- *Synonyms*—Synonyms are a data set containing user-defined synonyms. A default is provided.
- *Start List*—Start List is a data set containing the terms to keep. This is useful when technical jargon is part of an analysis.
- *Stop List*—Stop List is a data set of terms to be excluded.
- *Select Languages*—Select Languages defines the languages within the corpus to be processed. If blank, all languages will be processed.

Report Properties
- *Number of terms to display*—Number of terms to display defines the maximum number of terms that will be included in the output. Acceptable values are 500–20,000 or All.

When we parse the corpus of documents for the automobile claim fraud case (Fig. 9.8), the terms table can be useful for analyzing predicators of potentially fraudulent claims.

The term "fraud" appears in ten documents. This indicates that there are claims that the claim adjuster noted as potential fraud and would therefore benefit from further investigation. Parsing therefore can be a quick method for evaluating a large corpus to focus on specific terms that are the subject of analysis.

9.4 Zipf's Law

In the terms tables example (Fig. 9.8), notice that a small number of tokens makes up a relatively large distribution of the words from the corpus of 100 documents. George Zipf (1949) observed that there is an inverse relationship between the

Term	Role	Attribute	Freq	# Docs	Keep	Parent/Child Status	Parent ID	Rank for Variable numdocs
+ claimant	...Noun	Alpha	51	46Y		+	262	1
vehicle	... Noun	Alpha	33	30Y			110	2
accident	...Noun	Alpha	18	18Y			199	3
car	...Noun	Alpha	18	18Y			171	3
+ strike	... Verb	Alpha	17	17Y	+		194	5
+ door	...Noun	Alpha	14	14Y	+		138	6
+ cause	... Verb	Alpha	13	13Y	+		71	7
left	...Adj	Alpha	12	12Y			282	8
+ injure	...Verb	Alpha	11	11Y	+		283	9
right	... Adj	Alpha	11	11Y			27	9
+ allege	...Verb	Alpha	11	10Y	+		58	11
+ be	... Verb	Alpha	13	10N	+		413	11
fraud	...Noun	Alpha	10	10Y			174	11
+ hit	...Verb	Alpha	10	10Y	+		241	11
+ drive	... Verb	Alpha	9	9Y	+		277	15
truck	...Noun	Alpha	9	9Y			119	15
back	...Adj	Alpha	8	8Y			208	17
+ finger	...Noun	Alpha	10	8Y	+		222	17

Fig. 9.8 Text parsing terms table

frequency that a word is used within a corpus of documents and its rank within the frequency table. This is an approximation relationship that purports the most frequent word will occur approximately twice as often as the second most frequent, three times the third most frequent, etc. This law has also been studied in the context of other human behaviors and seems to be a useful approximation when the sample sizes are relatively small. In the realm of big data, the wider dispersions occur within the approximations. Expressed as an equation, Zipf's law states

$$f \propto 1/r$$

where f is the frequency of the occurrence of the word and r is the word rank. In other words, the product of the frequency of the occurrence and the rank is approximately constant. This approximation is useful for filtering terms when evaluating text. Often times we are neither interested in the most frequently or least frequently occurring terms. It is the terms in between that are the subject of interest.

9.5   Text Filter

Term reduction begins with a text filter (Fig. 9.9). Whereas text parsing provides a comprehensive list of terms from a large corpus of documents, text filtering enables focus on a specific term or terms.

Prior to applying a filter, the entire corpus is the subject of analysis. The filter then creates subsets for further analysis. The text filter can be used to identify new potential variables to provide a stronger relationship within a predictive model. Additionally, when building an initial predictive model, the text filter can be used to strengthen the training set by providing further verification of the accuracy of the target variable. In SAS Enterprise Miner™, text filtering enables rare terms to be eliminated. Further, an optional spell-check may further combine similar terms that were missed as a result of a misspelling. Filtering can be performed on a specific term or on a term along with its synonyms. Filtering is also used to set the term weight and frequency weight settings. When we apply Zipf's law to a corpus of documents, although we see a distribution that reflects both very high- and low-frequency terms, in both cases these are generally not useful for text analysis. Term discriminators that will be neither too high nor too low can be identified with the following document frequency:

$$\left(\frac{n}{100} \right) \geq \text{document frequency} \leq \left(\frac{n}{10} \right)$$

Fig. 9.9   Text filter

Frequency weight (Fig. 9.10) is a measure of the number of times that a term occurs across the corpus of documents. The frequency weighting property supports three methods to calculate frequency weight within SAS Enterprise Miner™. The *log* method is used to control high-frequency terms within a document. This is the default method. The *binary* method does not distinguish between high- and low-frequency terms; it is a measure of the presence or absence of the term. *None* is the setting that indicates that no transformation will take place (it does not indicate that no frequency weight will be calculated).

Term weights use one of three methods to determine the importance of a term. *Entropy* is a method that assigns weights between 0 and 1. Terms that occur infrequently will have higher weights. *Inverse document frequency* calculates the importance of a term as the inverse of the frequency of occurrence of the term in the corpus of documents. *Mutual information* is a calculation of differences in entropy. Mutual information provides insight on the extent to which a term belongs to a category. Term weights can also be set to *none*, indicating no term weights are applied.

Let's examine the key properties of the text filter node:

Fig. 9.10 Text filter properties

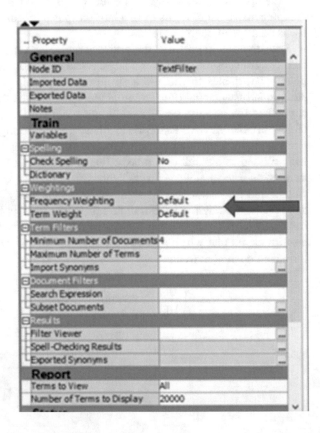

Train Properties

- *Check Spelling*—Check Spelling is a Yes/No indicator that identifies if the spelling checker is applied to the corpus. If Yes is specified but no dictionary is provided, then the lowest 5% of terms (based upon frequency) are considered candidates for misspelling, and the top 95% of terms are used as the dictionary.
- *Dictionary*—A Dictionary of terms to be used when Check Spelling is set to Yes.
- *Frequency Weighting*—Frequency Weighting uses log, binary, or none (described above).
- *Term Weight*—Term Weight uses entropy, inverse document frequency, mutual information, or none (described above).
- *Minimum Number of Documents*—The Minimum Number of Documents identifies the minimum document requirement for a term to be included.
- *Maximum Number of Terms*—The Maximum Number of Terms identifies the largest number of terms to be included.
- *Import Synonyms*—Import Synonyms is a data set that includes user-defined synonyms for stemming.
- *Filter Viewer*—To access the filter viewer, the node must have been executed at least once. The filter viewer enables searches on terms. If ># is specified before the term, then the search term and any synonym are used for the query.
- *Terms to View*—To indicate which terms will be displayed in the results, there are three options. *Selected* will display the terms with Keep = Yes. *Filtered* will display the terms with Keep = No. *All* will display the entire set of terms.
- *Number of Terms to Display*—The length of the list of terms to be displayed is set to either 500, 1000, 2000, 5000, 10,000, 20,000, or All.

If we apply a filter to the automobile claim fraud notes and specify >#fraud as the search term (Fig. 9.11), there are ten documents that are filtered. The TEXTFILTER_ RELEVANCE ranks the documents relative to the query term. In this case there was an exact match. Through the use of text parsing and text filtering, unstructured text can be converted to structured text for use within predictive analytics models.

9.6   Text Cluster

Cluster analysis (Chap. 8) is an unsupervised data mining technique used to find groupings within data. Cluster analysis differs from predictive modeling techniques because there is no target variable. A target variable creates a way to group a set of data around values contained within the target. In the absence of a target, various algorithms can be used to find associations within data. This concept applies to text data as well. In SAS Enterprise Miner™, text clustering is implemented through the use of a text cluster node (Fig. 9.12).

Clustering text data attempts to put documents that are similar into the same group of documents. Thus, documents that are dissimilar are therefore in different clusters. There are three techniques used to define similarity among documents:

| | | Search : | >#fraud | | Apply | Clear |

TEXT	TEXTFILTER_SNIPPET	TEXTFILTER_RELEVANCE
Claimant strained back from truck accident. refer to fraud unit.	... . refer to **fraud** unit. ...	1.0
Insured states he was driving and another person hit him causing him to strain his back.refer	... .refer to **fraud** unit. ...	1.0
Claimant loading car when insured vehicle struck him in the head ---- refer to fraud unit.	... - refer to **fraud** unit. ...	1.0
Claimant was lifting vehicle and alleges mid back pain. Closed file sent	... sent hq. **Fraud**? ...	1.0
Claimant injured her chin as a result of collision. Three months since accident. Failed to keep	... appointments. Possible **fraud** ...	1.0
Claimant injured her chin as a result of collision. Three months since accident. Failed to keep	... appointments. Possible **fraud** ...	1.0
Claimant alleges that he injured his right knee as a result of collision. Three weeks since	... doctors appointments. **Fraud**? ...	1.0
Claimant strained neck from truck accident. Refer to fraud unit.	... . Refer to **fraud** unit. ...	1.0
Sore back. Cause unknown. Possible fraud. Refer for investigation.	... unknown. Possible **fraud**. Refer for	1.0
Claimant rear-ended by another vehicle. Vehicle left the scene. Possible fraud.	... scene. Possible **fraud**. ...	1.0

Fig. 9.11 Filter results

Fig. 9.12 Text cluster

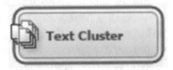

distance, correlation, and association. Mahalanobis distance (see Chap. 8) and Euclidean distance (see Chap. 3) are two of the most common methods for calculating distances between numbers. Mahalanobis distance is used within text clustering to calculate distance among documents. The magnitude of the distance determines if a document is considered similar or dissimilar. If correlation is used as the measure, then a high correlation indicates similarity, and a low correlation indicates dissimilarity. Association-based measures work well with a relatively small corpus of documents.

Once the method for determining how the documents relate is established, then clustering algorithms can be used to determine which documents belong within each cluster. There are two algorithms available in SAS Enterprise Miner™ to support text clustering. The algorithms are expectation-maximization (EM) and hierarchical clustering (Fig. 9.13). The expectation-maximization algorithm is a two-step iterative algorithm to find the maximum likelihood estimate. The algorithm assumes a probability density function for each document to belong to each cluster. The first step in the algorithm (expectation) creates a log-likelihood estimate of the current parameters using the *k-means* technique. Each document is given a weight (estimate) for each cluster. The second step (maximization) is an optimization step that maximizes the log-likelihood estimate from the expectation step by using the weights established in the expectation step. This step is repeated until the log-likelihood estimates do not change significantly. SAS Enterprise Miner™ uses the standard version (all of the data is considered within each step) of the EM algorithm for a small corpus. For a large corpus, the scaled version is used. The scaled version is a portion of the data within each iteration.

The hierarchical clustering algorithm uses Ward's minimum variance method (see Chap. 8) to calculate distances between clusters. This is an agglomerative technique where the criterion for choosing the pair of clusters is based on the optimal value of the lack-of-fit sum of squares.

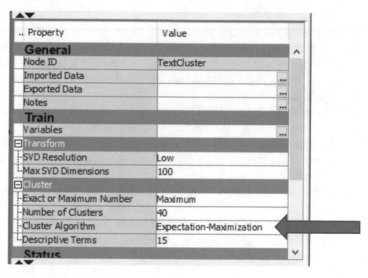

Fig. 9.13 Text cluster properties

Clustering occurs by evaluating the terms by document matrix. However, in a large corpus of documents, the terms matrix results in a very large matrix (i.e., high dimensionality). The more terms that you add, the more difficult it becomes to evaluate them; this is referred to as the *curse of dimensionality*. It is therefore a goal to reduce dimensionality. In Chap. 3, we presented one method of dimensionality reduction: principal component analysis. In statistics, confirmatory factor analysis is another commonly used method for dimension reduction. SAS Enterprise Miner™ uses *latent semantic indexing* for dimensionality reduction through *singular value decomposition* (SVD). Singular value decomposition is an eigendecomposition that breaks down the corpus of documents into linearly independent sets. In a large corpus, this will often result in many small sets that can be ignored, with the remaining sets becoming the subject of interest for analysis.

SVD computes three matrices (U, Σ, and V) from a term-by-document matrix A (represented as m by n), where m is the number of terms and n is the number of documents. The original matrix A can be expressed as A = UΣV, where:

1. U is the m by n matrix.
2. Σ is the rectangular diagonal matrix. The values on the diagonal are positive real numbers.
3. V is the n by n matrix.

By focusing on large eigenvalues, the result is a means to reduce terms in a large term by document matrix. This can result in a significant loss of terms; however, it produces a more meaningful subset of useful terms.

Let's examine the key properties of the text cluster node.

Train Properties

- *SVD Resolution*—SVD Resolution supports three types for SVD calculation. They are low, medium, or high resolution. High-resolution results in less loss of dimensionality but results in a greater number of terms to consider for analysis. Low resolution results in the greatest loss of information but the highest reduction in dimensionality.
- *Max SVD Dimensions*—Max SVD Dimensions specifies the maximum number of dimensions; the default is 100 and the maximum is 500.
- *Exact or Maximum Number*—Exact or Maximum Number identifies whether a specific number of clusters are to be created or if the maximum number of clusters will be created.
- *Number of Clusters*—Number of Clusters identifies the number of clusters to create when exact number is specified; the maximum is 999.
- *Cluster Algorithm*—The Cluster Algorithm specifies whether expectation-maximization or hierarchical clustering (described above) will be used.
- *Descriptive Terms*—For each cluster, Descriptive Terms states the number of terms used to describe the cluster.

When we apply a text cluster to the claim automobile insurance fraud example and change the exact or maximum number of clusters to exact with the number of clusters set to 4 (Fig. 9.14), we can then examine the results. Four clusters of unequal size are created (Fig. 9.15). The documents that are within cluster 2 contain all the documents for which claim adjusters included notes that either indicated the claim was fraudulent or potentially fraudulent. This is an example where the cluster terms (Fig. 9.16) have been reduced to a subset that provides more meaningful analysis enabling focus on the cluster(s) of interest based upon the descriptive terms. The number of clusters generated and the composition of the clusters are impacted by the algorithm used and the level of the SVD resolution. It may require multiple iterations to develop a cluster set that is appropriate for the subject.

9.7   Text Topic

In SAS Enterprise Miner™, the text topic node (Fig. 9.17) offers an alternative to the text cluster node for grouping documents. Within the text cluster, a document can appear in one and only one cluster. Within the text topic however, documents can appear in one or more topic groups based upon the relevance of the terms within the document. The text topic node creates topics of interest for analysis based upon groups of terms. The groups can be defined as single-term groups or multi-term groups. Like the text cluster node, the text topic node uses SVD to align the terms to document matrix for document grouping.

To utilize the text topic node, you must first have performed a text parsing and/or text filtering. The advantage of utilizing the text topic node after topic filtering is that the term weighting properties within the text filter will be used. If the text topic

Fig. 9.14 Text cluster
properties example

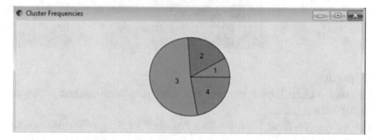

Fig. 9.15   Cluster frequencies

Cluster ID	Descriptive Terms		Frequency	Percentage
1	rearended +stop passenger vehicle +insure +state +claimant		8	8%
2	+fail collision 'possible fraud' +appointment keep three fraud +injure +cause 'fraud unit' unit +allege +hit +back +insure		18	18%
3	+finger 'car door' ankle +door left +injury +strain insured truck knee accident +strike +wrist back car		52	52%
4	cut hand +slip +back +park foot strain right customer passenger laceration +insure +drive pain +allege		22	22%

Fig. 9.16   Cluster terms

is used only after parsing the text, then weights will default to entropy for term
weighting and log for frequency weighting to determine which terms are grouped
within the topics. Once the node has been executed, then the topic viewer property
can be accessed to interactively evaluate topics.

Let's examine the key properties of the text topic node (Fig. 9.18).

Fig. 9.17 Text topic

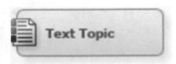

Fig. 9.18 Text topic
properties

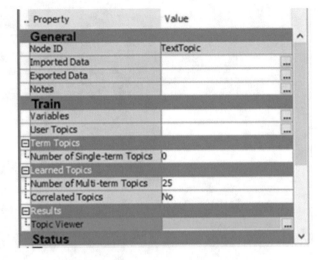

Train Properties

- *User Topics*—User Topics enables the use of a user-defined topic-term table (optional) for analysis.
- *Number of Single-Term Topics*—The Number of Single-Term Topics specifies the number of single-term topics to be created for analysis within the top weighted terms. The range is between 0 and 1000 (or the number of terms contained within the corpus of documents if it is less than 1000). The default is 0. Term weighting is used to create single-term topics (not SVDs).
- *Number of Multi-term Topics*—The Number of Multi-term Topics specifies the number of multi-term topics to be created for analysis within the top weighted terms. The range is between 25 and 1000 (or the number of terms contained within the corpus of documents if less than 1000). The default is 25. SVDs are calculated for multi-term topics.
- *Correlated Topics*—Correlated Topics specifies whether topics should be correlated.
- *Topic Viewer*—The Topic Viewer enables access to the interactive topic viewer. This is used to allow changes to topics by adding or removing terms (described below).

The text topic node results in two tables that are useful in understanding pertinent terms within a large corpus of documents. These include the topics table (Fig. 9.19) and the terms table (Fig. 9.20). The topics table identifies the specific terms that comprise each topic, including the number of terms and documents. The terms table provides a cross reference by identifying how each term weight applies to each topic to illustrate each topic that a term applied.

In addition, once the text topic has been executed, the interactive topic viewer (Fig. 9.21) enables an analyst to modify the terms that comprise topics and immediately evaluate the results. The term cutoff is a topic weight that measures the strength of the association between the term and the topic. Terms that are greater than or equal to the cutoff appear in yellow within the terms windows, while terms that are less than the topic weight appear in gray. Once the topics are finalized, it is possible to save the topics table for use in other projects. Topics can provide additional support within a predictive analytics project to identify variables that explain a target.

9.8   Text Rule Builder

The text rule builder node in SAS Enterprise Miner™ (Fig. 9.22) provides a predictive modeling technique for evaluating a categorical target with one or more text input variables. The text rule builder produces a set of rules that help explain a target variable and can subsequently be used when scoring new data for predictive analysis. Like the text cluster and text topics, the text rule builder uses subsets of the terms within a corpus of documents to build a set of Boolean rules. As a result, it must be preceded by text parsing and filtering. Since it requires a categorical target variable, it is a supervised learning technique. A data partition must be used prior to text parsing, and the target variable must be the same for the training, validation, and test data sets.

Let's examine the key properties of the text rule builder node (Fig. 9.23).

Train Properties
The train properties (generalization error, purity of rules, and exhaustiveness) contain the values very low, low, medium, high, and very high. In each case the default is medium.

- *Generalization Error*—The Generalization Error specifies that the validation data set is used to prevent overtraining. It is used to determine the predicted probabilities of the rules created from the validation data set. The higher the generalization error, the lower the probability of overtraining. However, higher

Category	Topic ID	Document Cutoff	Term Cutoff	Topic	Number of Terms	# Docs
Multiple	1	0.257		0.194 +injury,insured,+cause,+wrist,accident	7	12
Multiple	2	0.315		0.186 +finger,+door,right,car door,car	5	12
Multiple	3	0.286		0.194 three,collision,+injure,+fall,keep	9	7
Multiple	4	0.301		0.183 rearended,vehicle,passenger,+stop,+c..	5	8
Multiple	5	0.263		0.195 +strain,neck,unit,fraud unit,truck	9	10
Multiple	6	0.225		0.190 out,hand,right,laceration,right	5	8
Multiple	7	0.238		0.192 strain,back,+back,+strike,+park	4	11
Multiple	8	0.213		0.195 customer,foot,+drive,+allege,+hit	5	9
Multiple	9	0.247		0.194 left,truck,ankle,pain,knee	8	14
Multiple	10	0.253		0.195 +state,+insure,+cause,car,+slip	7	12

Fig. 9.19 Topics table output

Term	Role	Attribute	WEIGHT	Freq	# Docs	Keep	Rank for Variable NUMDOCS	+injury,insur ed,+cause,+ wrist,accide nt	+finger,+doo r,right,car door,car	three,collisio n,+injure,+fai l,keep	rearended,v ehicle,passe nger,+stop,+ claimant	+strain,neck, unit,fraud unit,truck	cut,hand,righ t,laceration,ri ght	strain,back,+ back,+strike, +park	customer,fo ot,+drive,+all ege,+hit
+claimant	Noun	Alpha	0.176158	51	46	Y	1	0.016	0.011	0.099	0.227	0.122	0.142	0.116	0.153
vehicle	Noun	Alpha	0.266509	33	30	Y	2	0.022	-0.038	-0.008	0.458	0.118	0.016	0.178	0.156
accident	Noun	Alpha	0.370991	18	18	Y	3	0.226	-0.004	0.226	0.028	0.195	0.083	0.004	-0.171
car	Noun	Alpha	0.370991	18	18	Y	3	-0.044	0.339	-0.044	-0.051	0.058	-0.019	0.081	0.048
+strike	Verb	Alpha	0.38343	17	17	Y	5	0.006	0.161	-0.051	0.018	0.056	-0.036	0.297	0.08
+door	Noun	Alpha	0.425683	14	14	Y	6	0.025	0.499	-0.066	-0.016	-0.024	-0.02	0.007	-0.021
+cause	Verb	Alpha	0.44181	13	13	Y	7	0.267	-0.04	-0.001	-0.028	-0.059	0.043	0.046	0.139
left	Adj	Alpha	0.459229	12	12	Y	8	0.21	0.043	-0.108	-0.016	0.001	-0.062	-0.053	-0.031
+injure	Verb	Alpha	0.478165	11	11	Y	9	0.193	-0.014	0.337	-0.02	-0.05	-0.048	0.007	0.047
right	Adj	Alpha	0.478165	11	11	Y	9	-0.037	0.339	0.142	0.009	-0.045	0.028	-0.012	0.145
+allege	Verb	Alpha	0.505591	11	10	Y	11	-0.072	-0.082	0.117	0	-0.191	0.066	0.13	0.327
+hit	Verb	Alpha	0.498906	10	10	Y	11	0.191	-0.011	-0.002	-0.03	0.13	-0.001	0.17	0.266
fraud	Noun	Alpha	0.498906	10	10	Y	11	-0.115	-0.022	0.256	0.006	0.261	0.009	0.006	-0.027
+drive	Verb	Alpha	0.521835	9	9	Y	14	-0.032	0.008	-0.037	-0.023	0.266	-0.039	-0.042	0.341
truck	Noun	Alpha	0.521835	9	9	Y	14	-0.142	-0.033	-0.063	-0.003	0.305	0.088	0.104	-0.13

Fig. 9.20 Term table output

Topics

Topic	Category	Term Cutoff	Document Cutoff	Number of Terms	# Docs
+injury,insured,+cause,+wrist,accident	Multiple	0.194	0.257	7	12
+finger,+door,right,car door,car	Multiple	0.186	0.315	5	12
three,collision,+injure,+fail,keep	Multiple	0.194	0.286	9	7
rearended,vehicle,passenger,+stop,+claimant	Multiple	0.183	0.301	5	8
+strain,neck,unit,fraud unit,truck	Multiple	0.195	0.263	9	10
cut,hand,right,laceration,right	Multiple	0.19	0.225	5	8
strain,back,+back,+strike,+park	Multiple	0.192	0.238	4	11
customer,foot,+drive,+allege,+hit	Multiple	0.195	0.213	5	9

Terms

Topic Weight	+	Term	Role	# Docs	Freq
0.532	+	injury	Noun	8	8
0.505		insured	Noun	8	8
0.267	+	cause	Verb	13	13
0.246	+	wrist	Noun	4	4
0.226		accident	Noun	18	18
0.217		neck	Verb	6	6
0.21		left	Adj	12	12
0.193	+	injure	Verb	11	11

Fig. 9.21 Topic viewer results

generalization errors have a higher probability of omitting potentially useful rules.

- *Purity of Rules*—The Purity of Rules controls the number of rules that will be created by adjusting the maximum p-value necessary for a rule to be created. The higher the value, the fewer the rules that will be created. Lower values will result in a greater number of rules created. Very high results in the fewest rules created and will contain fewer terms within those rules. These are considered purer rules. The p-values for each setting are:

1. Very low: $p < 0.17$
2. Low: $p < 0.05$
3. Medium: $p < 0.005$
4. High: $p < 0.0005$
5. Very high: $p < 0.00005$

- *Exhaustiveness*—Exhaustiveness determines how many potential rules are considered. A higher value will take longer to run and result in a greater number of

Fig. 9.22 Text rule builder

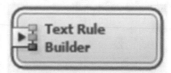

Fig. 9.23 Text rule builder
properties

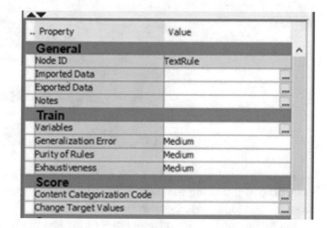

rules being created. This controls how many terms are tested. A higher value also
has a greater potential for overtraining.

Score Properties
- *Content Categorization*—Content Categorization is available after the node has
  been executed at least once. It shows the SAS content code that is generated for
  each target level.
- *Change Target Levels*—Change Target Levels is available after the node has
  been executed at least once. It allows you to view and modify the target levels.
  For changes to be applied, the node then needs to be rerun. All possible values
  (from the training, validation, and test partitions) are considered. This allows for
  interactive modeling of text rules.

Since the rules that are created are Boolean, they are easy to evaluate both a positive and a negative outcome. This can be used in evaluating a predictive model,
particularly with a binary target for evaluating scoring of a target.

9.9   Text Profile

The text profile node (Fig. 9.24) uses terms found within a corpus of documents to
profile a target variable. To maximize the possibility of creating discriminating
terms, terms that appear in more than one category have their term weights down-
weighed. The algorithm uses a Bayesian model to predict the terms that most likely
describe the category.

Fig. 9.24 Text profile

Fig. 9.25 Text profile
properties

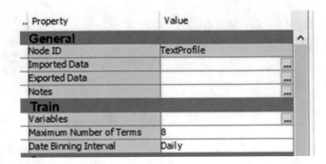

By downweighing the terms that appear in multiple categories, the algorithm goes beyond just using the most frequently occurring terms. In this way it differs from the text topic node (Sect. 9.7).

Let's examine the two key properties of the text profile node (Fig. 9.25).

Train Properties

- *Maximum Number of Terms*—Maximum Number of Terms defines the maximum number of terms that will be contained within each category. The default is 8 and 100 is the maximum.
- *Data Binning Interval*—The Data Binning Interval property is only applicable if the data contains a numeric variable with the role of Time ID. When present, it defines the period to bin the data within the entire corpus. The values include hourly, daily, weekly, monthly, and yearly. The default is daily.

The text profile provides an alternative method to strengthen a predictive model by aligning terms to specific categories. In this way qualitative data can be used to strengthen associations or correlations found within other predictive modeling techniques. Belief scores are calculated, and the terms with the highest beliefs are displayed. This can provide an indication of the independence of the terms. The text profile differs from the text topic because of the downweighing of terms that appear in multiple categories.

9.10   Chapter Summary

Qualitative (text) data has seen a tremendous growth in business analytics in recent years as organizations have looked to take advantage of numerous social media platforms around the world to promote and better understand their products and services. The implementation of big data repositories has contributed to the need and emphasis to utilize text mining as a method to support predictive analytics. There are many applications of the use of text mining including customer retention, analyzing competitors, spam filtering, and fraud detection. Text mining can be a stand-alone process or be utilized with quantitative data.

Text data can exist in many different file formats. In addition, *web crawling* and *web scrapping* make it possible to create large repositories of text data from one or more websites. It is possible to control both the depth and breadth of the data that is extracted from the web. Regardless of the source or file type, text repositories are referred to as a *corpus* of documents. The corpus serves as the input data source for text mining.

Parsing is used to reduce noise in text data by analyzing sentences into their logical syntactic structures. It is used to break sentences into individual terms that separate words into those that provide context and meaning. There are two approaches to parsing: *linguistics* and the *bag of words* method. Linguistics considers context and meaning. The bag of words method considers each word as an individual token. Parsing enables *stemming*, which is the ability to combine similar words such as verb tenses.

Zipf's law can be applied to a text filter to help further reduce a large number of terms into a smaller, more useful subset for analysis. In a large corpus of documents, we generally want to eliminate both very frequently occurring terms and very low frequently occurring terms. Term weights can be assigned to determine the importance of a term, eliminating those terms with low weights.

Text clustering is an unsupervised machine learning technique to group text data by evaluating similarity based upon distance, association, or correlation. Within SAS Enterprise Miner™, Ward's minimum variance or the expectation-maximization algorithm is used to calculate the distance between clusters. The number of clusters is specified by the analyst or determined by the algorithm. Clustering occurs by evaluating the number of terms in a corpus. In a large corpus, it is necessary to reduce the number of terms. This is accomplished through singular value decomposition. In a text cluster, a document can occur in one and only one cluster.

In SAS Enterprise Miner™, the text topic node offers an alternative to text clustering. Text is grouped by single-term or multi-term topics and the same document may occur in more than one topic. The text topic node uses either term weighting or entropy to determine which topic(s) a document belongs. The interactive topic viewer can be used to dynamically create topics and immediately evaluate the results.

The text rule builder node in SAS Enterprise Miner™ provides a predictive modeling technique for evaluating a categorical target with one or more text input variables. A set of Boolean rules are produced that can be subsequently used to score new data to help explain the target variable.

The text profile node is an alternative to the text topic node. It utilizes a Bayesian model to predict the terms most likely to describe a category. The model downweighs term frequency to maximize the possibility of creating discriminating terms. The text profile node can be used to strengthen associations or correlations found within other predictive modeling techniques.

Discussion Questions
1. What is meant by the term *text mining*?
2. Describe issues to consider when analyzing a large corpus of documents. Describe techniques that have to be utilized to overcome the issues.
3. Describe the role of each node when text mining using SAS Enterprise Miner™.
4. Describe an example of how filters can be used to analyze a large corpus of documents.
5. Describe methods that can be used to reduce dimensionality.
6. Explain how text mining can support predictive analytics.

References

Abbott D (2014) Applied predictive analytics: principals and techniques for the professional data analyst. Wiley & Sons, Indianapolis

Chakraborty G, Pagolu M, Garta S (2013) Text mining and analysis: practical methods, examples, and case studies using SAS®. SAS Institute Inc., Cary

Matthews K (2018) Here's how much big data companies make on the internet. Big Data Showcase. https://bigdatashowcase.com/how-much-big-data-companies-make-on-internet/

Zipf G (1949) Human behavior and the principle of least effort: an introduction to human ecology. Addison-Wesley, Cambridge, MA

Appendix A: Data Dictionary for the Automobile Insurance Claim Fraud Data Example

Attribute	Role	Level	Definition
Claimant _Number	ID	Interval	Unique identifier assigned to each claim
State_Code	Input	Nominal	Two-letter state abbreviation where the claim occurred
State	Reject	Nominal	Name of the state where the claim occurred
Claim_Amount	Input	Interval	Total amount paid for the claim
Education	Input	Nominal	Level of education attained by claimant (High School or Below, College, Bachelor, Master, Doctorate)
Employment_ Status	Input	Nominal	Employment status of the claimant (Employed, Unemployed, Medical Leave, Disability, Retired)
Gender	Input	Binary	Code indicating the claimant's gender (F, M)
Income	Input	Interval	Annual income of the claimant (in USD)
Marital_Status	Input	Nominal	Marital status of the claimant (Divorced, Married, Single)
Vehicle_Class	Input	Nominal	Type of automobile damaged as a result of the claim incident (Two-Door Car, Four-Door Car, Luxury, SUV, Luxury SUV, Sports Car)
Vehicle_Size	Input	Nominal	Category indicating the size of the vehicle that was damaged (Compact, Midsize, Luxury)
Fraudulent_ Claim	Target (dependent)	Binary	Code indicating if the claim was fraudulent (Y/N)

© The Author(s), under exclusive license to Springer Nature Switzerland AG 2022
R. V. McCarthy et al., *Applying Predictive Analytics*,
https://doi.org/10.1007/978-3-030-83070-0

Appendix B: Can You Predict the Money Laundering Cases?[1]

Introduction

This case is designed to be used in a business analytics course with a focus on the development and subsequent optimization of predictive models. Six of the nine steps in the predictive analytics process model (see Fig. B.1) (McCarthy et al. 2019) are reinforced through this case. The case begins with the development of a hypothesis that supports the business problem that is described herein. Data is provided with the case. The data must then be analyzed and manipulated for analysis. Predictive models are then developed and subsequently optimized to determine which model provides the best fit. The case provides an opportunity to build multiple predictive models using a variety of tools (e.g., SAS, R, Python) and to optimize those models. Once a best-fit model has been selected, it can then be implemented. One of the requirements for selection of the best-fit model is to produce a model that exceeds the benchmark within this case. A scoring data set is provided to analyze how the model supports ongoing production needs and to emphasize that the purpose of predictive models is to apply them to future business activities/decisions.

Students are provided with the background of the business problem, a data set for building and testing predictive models, and a data set to score the best model. The business problem is to develop an optimized predictive model to determine which cases must be investigated for potential money laundering. Next, the details of the business case are presented: beginning with a definition of money laundering and why it is a critical issue for the banking industry.

---

[1] McCarthy R, Ceccucci W, McCarthy M, Sugurmar N (2021, February) Can you predict the money laundering cases? Inf Syst Educ J 19(1):16–23. Reprinted with permission.

Fig. B.1 Predictive
analytics process model
(McCarthy et al. 2019)

Money Laundering

Money laundering is the illegal process of concealing the origins of money obtained
illegally by passing it through a complex sequence of banking transfers or commercial transactions. It is an attempt to make a large sum of money obtained through
illegal activities to look legitimate through banking transactions (Dreyer 2011). The
term money laundering comes from this process of taking "dirty" money (i.e., money
from illegitimate, criminal activities) and transforming it into "clean" money (i.e.,
money that appears legitimate and cannot be traced back to the criminal activity).
Money laundering is estimated to be a one to two trillion-dollar problem and may
represent 2% to 5% of the gross domestic product of the entire world (Ruce 2011).

Dreyer (2011) describes three layers of money laundering:

1. Placement—moving the funds from the criminal activity that generated them
2. Layering—using complex financial transactions to disguise the funds
3. Integration—making the money available for subsequent use

Initially, money laundering was a tool used by drug dealers and racketeers. More
recently, it is also viewed as a tool used by terrorists to finance illicit activities.

Bank Secrecy Act

The Currency and Foreign Transaction Reporting Act (1970) placed a requirement
on banks in the United States to work with the US government to investigate money
laundering (this act is more commonly known as the Bank Secrecy Act (BSA)).
This act placed a requirement on banks to report transaction activity that the government considers useful in monitoring criminal and tax matters (Ruce 2011). The
major tool to perform this monitoring was the creation of the *Suspicious Activity
Report* (SAR). A SAR must be filed when a bank has knowledge of or suspects that
a financial transaction has occurred as a result of funds earned through illegal activities. This report is one of the primary tools used to combat money laundering. There
are millions of transactions that occur daily. The vast majority of these transactions

are legitimate. However, while money laundering involves a relatively small number of transactions compared to the total number, their consequences can be severe. Under-reporting of suspicious activity could therefore lead to an increase in money laundering activity. Conversely, over-reporting can result in the investigation of too many legitimate transactions making it more difficult to focus on those that should be scrutinized (Meltzer 1991). It is important therefore to balance both issues and still meet the requirements of the Bank Secrecy Act. A report must be filed when:

1. The transaction is designed to evade the requirements of the BSA.
2. The transaction has no apparent lawful purpose, and the bank has no knowledge of a legitimate business reason for why the customer would engage in the activity resulting from the transaction (Ruce 2011).

The Bank Secrecy Act required the reporting of cash transactions that exceeded $5,000. The threshold was subsequently changed to $10,000 in 1984. This resulted in attempts to circumvent this requirement by structuring multiple transactions that were each below the $10,000 threshold. This caused requirements to be put in place to identify attempts to subvert this threshold. To address this problem, in 1986, the Money Laundering Control Act added to this requirement by criminalizing money laundering. The act defines *specified unlawful activities* (SUAs). These include attempts to conceal the source, control, or ownership of funds (Salinger 2013).

As a result of the September 11, 2001, terrorist attacks on the United States, the *PATRIOT Act of 2001* was passed to strengthen the penalties for terrorist acts that occur either domestically or abroad. This included money laundering and terrorist financing. It enhanced the due diligence requirements and SAR reporting requirements (Sensenbrenner 2001).

Business Problem

To comply with the requirements of the Banking Secrecy Act and subsequent legislation, banks must have a system in place to identify suspicious activity that has the potential to involve money laundering or terrorist financing. The volume of legitimate banking transactions that they process on a daily basis makes it unrealistic to manually evaluate every transaction. Therefore, they have systems and controls in place to flag suspicious activity for further investigation. It is helpful to have develop a probability for each transaction that indicates the likelihood that a transaction is suspicious for prioritization purposes (Mehmet and Buchholtz 2014). The system generates alerts using automated rules. The alerts need to be reviewed and investigated by anti-money laundering (AML) analysts. The goal of the alert system (AML system) is to come as close as possible to identifying only those specific transactions that involve these illegal activities. If too many false alerts are generated, it consumes too much time on the part of the analysts to investigate each one. If too few alerts are generated, then there is the potential to miss transactions that support illegal activity. The better the model for evaluating alerts, the closer the system gets to its optimal performance.

When an analyst reviews an alert, either they create a case for further investigation because they suspect fraud or they close the alert. Cases requiring further investigation are sent to an operations team for review. Upon completion of the review, cases that still require further investigation require the filing of a Suspicious Activity Report (SAR). Alerts are reviewed based upon priority (i.e., the higher the probability of suspicious activity).

Analyze Data

The data for this case was provided by People's United Bank and consists of 38,515 transactions that were a sample of production transactions from October 2014 through September 2015. This data set was used to produce their model. The data set and accompanying documentation serves as the baseline to develop a predictive model that improves upon this baseline.

People's United Bank is a diversified financial services company with more than $60 billion in assets. Founded in 1842, People's United Bank is a community-based, regional northeast bank with more than 6500 employees, over 400 retail locations in Connecticut, New York, Massachusetts, Vermont, New Hampshire, and Maine. There are over 148 branch locations co-located within Stop & Shop grocery stores in Connecticut and New York. They offer full-service, extended-hour retail banking and commercial banking as well as wealth management services.

Two data sets were created from the dataset provided. The first data set is used to develop and train a predictive model to determine if a case requires further investigation for potential money laundering (i.e., the case should result in a SAR). The first data set contains 35,000 observations (records). The data dictionary is presented in Table B.1.

The second dataset used for scoring contains the remaining 3,515 observations. This dataset is used to test the best-fit model and determine differences in cases selected for further review between the best-fit model and the baseline model. The datasets are provided in an Excel, CSV, or SAS file format providing flexibility for a variety of analytic tool use.

Development and Optimization of a Best-Fit Model

The baseline model was developed by People's United Bank by analyzing six iterations of predictive models. The best-fit model was chosen using the Kolmogorov-Smirnov (KS) and receiver operating characteristic (ROC) statistics. The best-fit model was developed using 70% of the data to build the model and 30% to validate the model. The best-fit model for this baseline was a backward regression model. The statistics for the model are presented in Table B.2.

Table B.1 AML data dictionary

Variable	Type	Note
ID	Unique ID	Unique identifier for alerts generated by the anti-money laundering (AML) system
Wires_mult	Derived	The number of wire transfers with more than $10K If an alert has one-wire transfer more than $10K, then wires_mult will be 0 If the same alert has three-wire transfers with $10K transfer for each wire, then wires_mult will be 2
Wires_size	Derived	The total wire transfer involved in the AML alert is proportion to the size of the total amount =Sum (# of wire transfers)/10,000; rounded down to the nearest integer For example, if an AML alert has 25,000 wire transfers, wires_size will be 25,000/10,000 = 2
Max_crr	Direct	Maximum cash reserve ratio score For example, ID 123 has five scores available in scoring table—the largest value is used
Num_tran_alert	Derived	Number of distinct transactions involved in a specific AML alert
Num_acct_alert	Derived	Number of distinct accounts involved in a specific AML alert
Num_related	Derived	Number of related transactions—for that specific AML alert
Num_tran_type	Derived	Number of different transaction types involved in the AML alert
Num_tran_bin	Derived	Number of trigger transactions are grouped in bins by the following order: If num_tran_alert ≤ 5 = '1' 6–15 = '2' 16–25 = '3' 26–50 = '4' 51–100 = '5' ≥ 101 = '6'
Num_acct_bin	Derived	Number of accounts involved in the AML alert are grouped by the following order: If num_acct_alert ≤ 1 = '1' 2–2 = '2' 3–3 = '3' 4–4 = '4' ≥ 5 = '5'

(continued)

Table B.1 (continued)

Variable	Type	Note
Trig_amt_ bin	Derived	Trigger amount stratified in ranges: If the amount <15000 = '1' 15,000–50,000 = '2' 50,000–100,000 = '3' 100,000–250,000 = '4' 250,000–500,000 = '5' 500,000–1,000,000 = '6' 1,000,000–2,000,000 = '7' >2,000,000 = '8'
Rela_amt_ bin	Derived	Related transaction amount stratified in ranges: If the related amount <15000 = '1' 15,000–50,000 = '2' 50,000–100,000 = '3' 100,000–250,000 = '4' 250,000–500,000 = '5' 500,000–1,000,000 = '6' 1,000,000–2,000,000 = '7' >2,000,000 = '8'
Scen_Cat_ code	Derived	If an AML alert belongs to a certain scenario, then it is set to 1 else it is set to 0. This is computed by analyzing the last 12 months of AML alerts
Num_rela- bin	Derived	Number of related transactions are grouped in bins by the following order. value numfmt = '1' <5 = '1' 6–15 = '2' 16–25 = '3' 26–50 = '4' 51–100 = '5' 101> = '6'
Num_ trigger	Derived	Number of trigger transactions for that specific AML alert
Prod_ind	*Target*	Binary: 1. Indicates the alert is a productive alert. Productive is defined as that particular alert for further investigation for AML-related activities. 2. 0 indicates the alert is not a productive alert

Table B.2 Baseline model statistic

Statistic	Result
KS	44.13
ROC	0.7923
R-square	0.1687

Regression is one commonly used technique when developing predictive models; however many other techniques are available in the data analyst's toolbox. The goal is to develop a model that provides a better fit. Even a modest improvement can be significant as it will result in better control over the review of cases evaluating potentially illegal activities. Any predictive technique may be utilized; this is not limited to only using regression models.

Final Report

The best-fit model enables a bank to identify and detect potentially illegal activity more accurately and quickly. It provides a more precise review of transactions that require investigation and can reduce the investigators workload.

In your final report, you should discuss the following:

1. Determine hypotheses:

   What were the hypotheses that you tested? If any variables were excluded, discuss why they were removed from the subsequent analysis.

2. Analyze data

   Which variable(s) contained missing values and how were they treated?

   Why was the technique chosen appropriate?

   Which variable(s) contained outliers?

   How did you identify and handle them?

   What variable(s), if any, were skewed?

   How did you handle them?

   What partition size(s) were utilized?

   It is appropriate to replicate the 70/30 split that the baseline utilized; however, if a more optimal partition was utilized that did not result in over-training, the model then discusses the results.

3. Predictive model

   For each predictive model technique, discuss the technique and the properties that resulted in the best-fit model.

   Compare the results of the selection statistics, and discuss which model produced the best fit. Present the results of all of the selection statistics. Compare the misclassification rate of each model type that was evaluated, and discuss the difference in the results of both type I and type II errors. Recall that a type I error occurs when a true null hypothesis is rejected and a type II error occurs when you fail to reject a null hypothesis when it is really false. How do these errors impact investigators?

4. Scored results

   Which cases resulted in the five highest probabilities for generating a SAR? Were there any cases that were not previously flagged by the baseline model that were flagged by your best-fit model?

References

Dreyer T (2011, December) Anti-money laundering. Credit Union Management, 40–42

McCarthy R, McCarthy M, Ceccucci W, Halawi L (2019) Applying predictive analytics: finding value in data. Springer Publications

Mehmet M, Buchholtz M (2014) Money laundering detection framework to link disparate and evolving schemes. J Digit Forensics Secur Law 8(3):41–70

Meltzer P (1991) Keeping drug money from reaching the wash cycle: a guide to the banking secrecy act. Bank Law J 108(3):230–255

Ruce P (2011) The bank secrecy act: considerations for continuing banking relationships after the filing of a suspicious activity report. Quinnipiac Law Rev 30(1):43–68

Salinger LM (2013) Encyclopedia of white collar and corporate crime, vol 1(2). SAGE, Thousand Oaks, pp 83–84

Sensenbrenner J (2001) H.R.3162 – Uniting and strengthening America by providing appropriate tools required to intercept and obstruct terrorism (USA PATRIOT ACT) act of 2001. https://www.congress.gov/bill/107th-congress/house-bill/3162

Index

Printed in the United States
by Baker & Taylor Publisher Services